International Conference on the Use of X-Ray (and related) Techniques in Arts and Cultural Heritage 2011

(XTACH 11)

IOP Conference Series: Materials Science and Engineering Volume 37

Sharjah, United Arab Emirates
7 – 8 December 2011

ISBN: 978-1-63439-848-0
ISSN: 1742-6588

Printed from e-media with permission by:

Curran Associates, Inc.
57 Morehouse Lane
Red Hook, NY 12571

Some format issues inherent in the e-media version may also appear in this print version.

Printed by Curran Associates, Inc. (2015)

Institute of Physics
Dirac House, Temple Back
Bristol BS1 6BE UK

Phone: 44 1 17 929 7481
Fax: 44 1 17 920 0979

techtracking@iop.org

Additional copies of this publication are available from:

Curran Associates, Inc.
57 Morehouse Lane
Red Hook, NY 12571 USA
Phone: 845-758-0400
Fax: 845-758-2634
Email: curran@proceedings.com
Web: www.proceedings.com

TABLE OF CONTENTS

PREFACE

The International Conference on the Use of X-Ray (and related) Techniques in Arts and Cultural Heritage (XTACH11) was held on 7 and 8 December 2011 at the American University of Sharjah (AUS) in the United Arab Emirates.

The conference was organized in collaboration with the International Atomic Energy Agency (IAEA) and the National X-ray Fluorescence Laboratory (NXFL).

The conference was inaugurated by Dr Peter Heath, Chancellor of the American University of Shrjah and attended by Mr Kwaku Aning, deputy Director General of the International Atomic Energy and Ambassador Hamad Al-Kaabi, Ambassador of the UAE to the International Atomic Energy university officials, faculty and students. The conference covered a variety of topics including the use of x-ray and micro beam x-ray analysis, synchrotron based techniques, ion beam and neutron based techniques, optical imaging and mass spectroscopy and chromatography techniques as well as best conservation practices.

XTACH11 provided an excellent forum for scientists in the region to interact, exchange ideas and to initiate collaborations with each other as well as with the international community. It showcased some of the latest technical developments in the field of non-destructive testing for the diagnosis and conservation of cultural heritage materials. In addition to the presentations by the invited speakers (Rene van Grieken and K Janssens, University of Antwerp, Belgium; Thomas Calligaro, Centre de Recherche et de Restauration des Musees de France; Stefano Ridolfi, Ars Mensurae, Rome, Italy, and Andrzej Markowicz, IAEA, Austria), a total of 25 other research papers were also presented and discussed. Scientists from many countries participated in the conference: Austria, Belgium, Egypt, Italy, India, Iran, Iraq, Jordan, Lebanon, Qatar, Saudi Arabia, Spain, Sri Lanka, Syria, Thailand, the United Arab Emirates and Yemen.

The conference concluded with a Discussion Panel. Thomas Calligaro (Centre de Recherché et de Restauration des Musees de France) introduced the Cultural Heritage Advanced Research Infrastructure (CHARISMA) website, highlighting some of its components such as ARCHLAB Infrastructures (for bibliographies), FIXLAB (for large instruments) and MOLAB (for mobile instruments). He explained that CHARISMA was for people working in Europe (or in associated countries). There was some discussion then about the need for a similar sort of network being established for the Middle East region.

The proceedings contain peer reviewed papers presented at the conference.

During the conference and in collaboration with the Sharjah Department of Culture and Information (Directorate of Antiquities), the American University of Sharjah organized an archaeology exhibition from the 30 November until the 10 December 2011. Special thanks go to Dr Sabah Jasim and Eisa Abbas from the Sharjah Directorate of Antiquities for facilitating this event. The exhibition included about 45 artifacts from different sites within Sharjah emirate, some of which dated as far back 7000 years ago. These artifacts included: flint arrowheads, pottery and alabaster vessels, metal objects including bronze arrowheads and coins, as well as carnelian bead necklaces.

Also coinciding with the XTACH11 conference, the American University of Sharjah, in cooperation with The International Atomic Energy Agency (IAEA) and the National X-ray Fluorescence Laboratory organized a regional training course from 3–8 December 2011. This covered a similar range of topics to those discussed at the XTACH11 conference. Participants of the course were trained in the use of physical and analytical techniques in cultural heritage. In addition to lectures from the National X-ray Fluorescence Laboratory and from the IAEA, the NXFL team provided the opportunity for trainees to undertake projects on ancient pottery samples, metal artifacts and Islamic manuscripts. The practical part of the course included experiments on XRF (portable, and micro XRF), Raman measurements, as well as other sample preparation techniques. The training course concluded with a series of presentations of the results by the participants, attended by the NXFL team and experts from the IAEA. This training course was organized as part of the activities of the IAEA technical cooperation RAS1011 Project: Using Ion Beam Analysis and Complementary Nuclear Techniques for Material Characterization in ARASIA State Parties. The course was attended by participants from Iraq, Jordan, Lebanon, Qatar, Syria, the United Arab Emirates and Yemen.

We would like to thank the conference sponsors: Tashkeel, Sharjah Museums, and Sharjah Department of Culture and Information (Directorate on Antiquities). We would also like to thank our invited speakers, the international advisory committee, the referees and the participants. It has been a pleasure working with them all.

Organizing Committee

Nasser Hamdan AUS, NXFL

Members
Pia Anderson AUS
Hussain Al-Awadhi UoS & NXFL
Sami El Khatib AUS
Attaelmanan Gaffar UoS & NXFL
Johannes Giesen AUS
Sabah Jasim Directory of Antiquities Sharjah
Najeh Jisrawi UoS & NXFL
Adil Tamimi AUS

International Advisory Committee

Zaki Aslan ATHAR, ICCROM, Italy
Mark Beech Abu Dhabi Culture & Heritage, UAE
Rene Van Grieken University of Antwerp, Belgium
Gene Hall Rutgers, The State University of N.J.

Peter Jackson Office of the Ruler of Sharjah
Andreas Karydas IAEA Laboratories, Seiberdorf, Austria
Giacinto Porco The Italian Association of Non-Destructive Testing, Italy
Mohammad Roumie (CNRS) Lebanese Atomic Energy, Commission, Lebanon

Acknowledgments

The organizers gratefuly acknowledge the support received from the College of Arts and Sciences and the Provosts office at the American University of Sharjah. The support of the Physics Section, Division of Physical and Chemical Sciences at the IAEA is greatly appreciated. We would like to give special thanks to Dr Francoise Muelhauser from the IAEA. We would like also to thank the conference sponsors: Tashkeel, Sharjah Museums, and Sharjah Department of Culture and Information (Directorate on Antiquities). We thank our invited speakers, the international advisory committee, the referees and the participants.

Conference photograph.

Sponsors' logos

tashkeel

National X-ray Fluorescence Laboratory

المعمل الوطني للأشعة السينية

إدارة متاحف الشارقة

Sharjah Museums Department

متحف الشارقة للآثار

SHARJAH ARCHAEOLOGY MUSEUM

حكومة الشارقة . دائرة الثقافة والإعلام

Government of Sharjah

Department of Culture & Information

IOP Conf. Series: Materials Science and Engineering **37** (2012) 011002 doi:10.1088/1757-899X/37/1/011002

Andrzej Markowicz

Dr. Markowitz obtained his Ph.D in Applied Nuclear Physics from the Institute of Physics and Nuclear Techniques, Cracow, Poland in 1977 and Habilitation in Radiation Physics from the Institute of Physics and Nuclear Techniques, Cracow, Poland in 1988.

Dr. Markowicz is currently the Head, Nuclear Spectrometry and Applications Laboratory at the IAEA laboratories in Siebergdorf, Vienna since 1991.

Dr. Markowitz professional career covers a wide range of topics including the following topics:

- **radiation physics**
- **X-ray spectrometry** including development of X-ray fluorescence instruments for laboratory and in-situ applications, improvements in methodology, accuracy and extension of applicability range of XRF techniques
- **electron-probe microanalysis** (EPXMA) including development of new atomic number and X-ray absorption procedures and theoretical description of bremsstrahlung background in EPXMA of multi-element bulk and thin specimens
- **industrial applications** of on-line and in-situ XRF analyzers
- **developing reference materials** for trace element analysis
- **total reflection XRF and X-ray micro-fluorescence** analysis including methodology and applications
- practical applications of X-ray fluorescence technique for **environmental pollution monitoring and study of cultural heritage objects**
- **Quality Assurance and Quality Control** principles and procedures applied for analytical techniques, in particular to X-ray emission spectrometry

Dr. Markowicz has published around 95 Scientific papers, review articles and book chapters. He is a m ember of Advisory Board of X-Ray Spectrometry Journal, Co-editor (together with Prof. R.Van Grieken, Belgium) of Handbook of X-Ray Spectrometry, Methods and Techniques, Marcel Dekker, New York.

IOP Conf. Series: Materials Science and Engineering **37** (2012) 011002 doi:10.1088/1757-899X/37/1/011002

Review of the IAEA Activities in Support of Characterization and Properties of Cultural Heritages Artifacts

A. Markowicz, A. G. Karydas, R. Padilla-Alvarez

a.markowicz@iaea.org

Nuclear Spectrometry and Applications Laboratory, Physics Section, IAEA Laboratories, A-2444 Seibersdorf, Austria

One of the major areas of the applications of nuclear techniques such as nuclear analytical methods (XRF, IBA and NAA), dating techniques and gamma- and electron-irradiation methods is characterization, study and protection of cultural heritage (CH) objects. IAEA provides an extensive support to the laboratories in Member States in these fields through Technical Cooperation (TC) projects focused on establishing technical infrastructure and development of skilled personnel, Coordinated Research Projects (CRPs) designed for coordination of research and development in the laboratories of both the developed and developing countries, and applied research and development activities carried out in the IAEA Nuclear Spectrometry and Applications Laboratory (NSAL) at Seibersdorf. The recent results and selected applications of X-ray fluorescence (XRF) techniques for laboratory and in-situ measurements will be presented. Since XRF techniques provide information on the elemental composition and 2D/3D elemental distribution in the CH objects in a truly non-destructive manner these methods became extremely attractive and commonly applied by restorers and conservators. A broad list of possible investigations includes, inter alia, preventive conservation, determination of production technology and origin, evaluation of alteration processes and diagnosis of previous modifications. Major advantages of the XRF techniques are simplicity, generation of multi-element analytical data, speed of operation, portability and flexibility in terms of analysis of various objects. NSAL developed the appropriate facilities for characterization and study of CH objects such as laboratory and (trans) portable EDXRF spectrometers, micro-beam XRF and absorption techniques and SEM. Moreover, under a special agreement the access to IBA techniques and synchrotron radiation sources in the laboratories in Croatia and Germany was ensured. New applications of the XRF techniques for the elemental compositional characterization and 3D imaging for a wide range of CH objects like paintings, coins, sculptures, bronzes have been developed and implemented in cooperation with the Museum of Fine Arts in Vienna.

IOP Conf. Series: Materials Science and Engineering **37** (2012) 011002 doi:10.1088/1757-899X/37/1/011002

Koen Janssens

Koen Janssens is professor of general and analytical chemistry at the University of Antwerp in Belgium. He obtained his PhD in 1989 on a thesis dealing with the use of artificial intelligence techniques for automated treatment of X-ray analysis data. Since then, he has been actively making use of X-ray microbeams, produced in laboratory sources such as X-ray tubes or generated in large accelerator complexes called synchrotron storage ringe, for non-destructive materials analysis.

Next to devoting is attention to the development of methods for X-ray based materials analysis, he also employs synchrotron micro- and nanobeams for studying questions of environmental nature, mainly dealing with heavy metal pollution of natural materials such as soil and sediment. In the area of cultural heritage, he employs X-ray based methods to better understand alteration and degradation processes of different materials such as glass, inks and paintings. He is author of ca 150 scientific papers and edited several scientific books on (non-destructive) X-ray analysis and related topics.

IOP Conf. Series: Materials Science and Engineering **37** (2012) 011002 doi:10.1088/1757-899X/37/1/011002

Abstract

X-ray based imaging and spectroscopy of cultural heritage materials

K. Janssens,
Department of Chemistry, University of Antwerp, Universiteitsplein 1, B-2610 Antwerp, B

The mission of cultural heritage institutions is to preserve and protect artifacts from the distant or more recent past. In order to fulfill this mission in a professional manner, detailed knowledge on alteration phenomena of various kinds that gradually and unobtrusively are taking place at or below the surface of these objects is required.

In order to be able to investigate the nature of these chemical transformations that sometimes lead to the formation of microscopically thin alteration layers, the use of state-of-the-art micro analytical methods is required. Next to being able to provide information on the composition of various materials at or just below the surface, these methods also must be able to deliver highly specific information on the nature of the chemical compounds that are locally encountered. In this respect, our recent experience shows that the use of a combination of synchrotron X-ray based spectroscopic and imaging methods such as X-ray fluorescence analysis, X-ray absorption spectroscopy and X-ray diffraction can reveal significantly new information of certain alteration processes that have remained enigmatic for a long time. Concrete examples to be discussed is the darkening of originally yellow lead chromate paint layers, as encountered in paintings of V. Van Gogh and the blackening of red cinnabar-based paint layers in works of Rubens.

Since many decades, X-ray radiography (XRR) is a standard method employed in many cultural heritage institutions and musea to non-destructively inspect paintings. XRR can reveal hidden/overpainted layers in such artifacts, provided the absorption contrast is sufficiently large. By employing element-specific X-ray fluorescence and/or crystal-phase specific X-ray diffraction contrast instead of the X-ray absorption contrast, the information that can be recorded about hidden layers in a painting can be significantly enhanced. The possibilities and limitations of using macroscopic XRF and XRF/XRD scanning for investigation of paintings by artists such as Vincent van Gogh and Rembrandt van Rijn will be discussed.

IOP Conf. Series: Materials Science and Engineering **37** (2012) 011002 doi:10.1088/1757-899X/37/1/011002

Stefano Ridolfi
Ars Mensurae, Rome, Italy

Stefano Ridolfi is a physicist. He is CEO of Ars Mensurae in Rome, the first spin off company in Italy specifically devoted to the development and use of portable non-invasive systems for the analyses of Cultural Heritage. He teaches at University "Sapienza". He is author of several scientific papers and book chapters related to the optimization and use of portable ED-XRF systems applied to Cultural Heritage. His current field of research is the improvement of micro XRF portable systems for "in situ" analyses and the development of specific SOP's for ancient metals, polychrome surfaces and stone like materials.

IOP Conf. Series: Materials Science and Engineering **37** (2012) 011002 doi:10.1088/1757-899X/37/1/011002

Professor René Van Grieken

Professor René Van Grieken received a Master degree in Chemistry (1967) and a PhD in Sciences (1971) from University of Ghent in Belgium. **Professor René** received "Aggr HO" (comparable to "Habilitation" in Germany) from University of Antwerp in 1976.
Currently, Dr. René is an Emeritus Professor at the Department of Chemistry, University of Antwerp, Belgium.

Professor Van Grieken was co-founder (together with Freddy Adams, who retired in 2003, and Renaat Gijbels, who retired in 2004) of the Micro- and Trace Analysis Centre (MiTAC) in the Department of Chemistry at the University of Antwerp.

Professor Van Grieken career history is very comprehensive and full of fruitful productive research and academic work, and here we list few lines:

* 1967-73: Assistant, subsequently research fellow, then research associate of the Belgian National Science Foundation, at the Institute of Nuclear Sciences, **University of Ghent**

* 1972-73: Visiting professor at **Florida State University**, Tallahassee, FL., USA (research on the development of proton induced X-ray emission (PIXE) for quantitative analysis of atmospheric aerosols)

* 1973-85: "Docent" then "Hoogleraar", Department of Chemistry, University of Antwerp

* 1985-2010: Full Professor, Department of Chemistry, University of Antwerp (teaching instrumental analysis to Pharmacy and Chemistry; environmental chemistry to Biology, Chemistry, Environmental Sciences and Safety Engineers; radioprotection to Chemistry and Medicine)

Professor Van Grieken published 592 publications in international scientific journals with peer review on about micro- and trace analysis and applications (mostly instrumental and environmental analysis, and preventive conservation).

Professor Van Grieken is author and editor of 13 books on atmospheric aerosols, environmental analysis, X-ray spectrometry and mass spectrometry.

Professor Van Grieken has 221 chapters in books and contributions to proceedings of international conferences. Professor René work is quoted in about 1100 citations.

Professor René Van Grieken with ***h***-index of 48 full record and work history can be found at the University of Antwerp website at http://webh01.ua.ac.be/mitac1.

XTACH 11 IOP Publishing
IOP Conf. Series: Materials Science and Engineering **37** (2012) 011002 doi:10.1088/1757-899X/37/1/011002

IS CHEMICAL ANALYSIS NEEDED FOR CULTURAL HERITAGE STUDIES?

René Van Grieken, Department of Chemistry, University of Antwerp, B-2610 Antwerp, Belgium

Abstract: There are basically three reasons why chemical analysis is often invoked in the realm of cultural heritage (CH) studies:

-Our fundamental interest in the materials and methods used by the artists
-Tracing of forgeries, and elucidating the provenance of materials in CH
-Preparing for restorations, so that incompatible methods and materials are avoided.

X-ray spectrometry (XRS) is very important for CH studies, essentially because it is non-destructive. This appears e.g. from the program of a recent conference on non-destructive and micro-analytical techniques in art and CH: 115 papers were based on X-ray techniques (namely 60 on energy-dispersive X-ray fluorescence (ED-XRF), including portable and micro-XRF, 25 on SEM/EDX, 14 on XRD, 10 on PIXE, 3 on X-tomography, 2 on X-radiography), while in total 113 were based on all non X-ray techniques (e.g. 35 diverse optical techniques, 26 on (FT-)IR, 15 on Raman spectrometry, 10 on GC(-MS), etc.).
There are thousands of examples in the literature (but often in the "grey" literature) on the use of XRS for CH, ranging from very simple ones, to very complicated ones, using "common" laboratory XRF (WD and ED) if samples are small and can be brought to the lab (e.g. historical coins), portable XRF (radioisotope- and tube-excited), micro-XRF (lab-based or portable), synchrotron XRF (usually micro or confocal) and X-ray absorption near-edge spectrometry (XANES), total reflection XRF, particle induced X-ray emission analysis, and electron probe X-ray microanalysis.
"Conservation" implies, in its most simple definition: maintaining a work of art in «good condition». In addition to intrinsic parameters, often external environmental factors are important for conservation, even indoors. Improving the environmental conditions (micro-climate and chemical pollution) around a work of art is now popular and called: preventive conservation. Also here, XRS techniques have appeared to be very important.

IOP Conf. Series: Materials Science and Engineering **37** (2012) 011002 doi:10.1088/1757-899X/37/1/011002

Thomas Calligaro

Centre de Recherche et de Restauration des Musées de France (C2RMF), Paris, France

Dr. Thomas CALLIGARO is the Head of the *Mineral Materials* research group and permanent researcher at the *Centre de Recherche et de Restauration des Musées de France* (C2RMF), a joint unit of the French ministry of culture and of the *centre national de la recherche scientifique* (CNRS UMR 171). He obtained his PhD in high energy nuclear physics on " Design, test and use of a simulation program for a multiwire proportional chamber for the detection of photoelectrons " Delphi experiment – LEP - CERN Geneva, Switzerland in 1986.

His research interests are in Non-destructive and non-invasive characterization of materials of cultural heritage. Specific research programs applied to ancient ornamental stones and lithic tools from museum collections (rubies, emeralds, garnets, quartz, and obsidian, jade, turquoise). He also works on development of analytical methods based on radiations and experimental set-ups specifically tailored to address issues in the field of Art History and Archaeology (e.g. : in-air ERDA with a nuclear microprobe, PIXE-induced X-ray fluorescence, XRF with mobile instruments).Dr. Calligaro uses various Analytical techniques such as ion beams techniques (PIXE, PIGE, RBS, ERDA, NRA), XRF and Raman micro-spectrometry. He has a large number of publications in the field of non-destructive testing in art and archaeology.

IOP Conf. Series: Materials Science and Engineering **37** (2012) 011002 doi:10.1088/1757-899X/37/1/011002

Abstract:

The merits of Ion Beam Analysis in evidencing art and archaeological fakes

Thomas Calligaro
Centre de Recherche et de Restauration des Musées de France, CNRS UMR171
Palais du Louvre, 75001 PARIS - FRANCE
Detection of fakes and forgeries in art and archaeology represents a fundamental question. To address this issue, cultural heritage institutions request the expertise of scientific laboratories. There is a constant competition between, on one side, the forger and its skill, and on the other side, the knowledge of experts in authentic materials and the performance of the analytical techniques they use. Because investigations are conducted prior to a possible acquisition, non-destructive and non-invasive methods are mandatory.

Materials analysis using ion beams produced by accelerators, providing the chemical composition with excellent performance and total non-destructiveness, have many advantages. We present here cases studies of forgeries detection at the Centre de recherche et de restauration des musées de France using Ion Beam Analysis (IBA) techniques, namely PIXE , PIGE , RBS , NRA and ERDA with the AGLAE particle accelerator.

The detection of fake can rely on the simple identification of erroneous materials using PIXE, such anachronistic pigments in painted works, or by the presence of materials produced following a recipe not existing at the time the object was supposed to be manufactured, or through the invisible alteration layer at its surface by time.

The presentation will be illustrated through several examples: the PIXE study of the Vinland map, a controversial document supposedly representing the America 50 years before its discovery by Columbus, the PIXE and PIGE analysis of the Egyptian blue head of the Louvre museum which revealed its forgery nature and the evidence of a fake Pre-Columbian rock crystal skull though the hydrogen depth profiles obtained by ERDA. Eventually, the presence of both genuine and fake elements on specific artefacts can even question the notion of fake.

e-mail: thomas.calligaro@culture.gouv.fr

IOP Conf. Series: Materials Science and Engineering **37** (2012) 012001 doi:10.1088/1757-899X/37/1/012001

Portable X-ray Fluorescence Spectrometry for the analyses of Cultural Heritage

S Ridolfi
CEO, Ars Mensurae, Italy
stefano@arsmensurae.it

Abstract

Field Portable Energy Dispersive X Ray Fluorescence (FP-EDXRF) is particularly useful to analyze works of art, mainly because of his noninvasive and multielemental capability. In many situations FP-EDXRF is the only non invasive technique that can be realistically used to gain some information about the chemical composition of precious and unique objects. Many kind of works, such as paintings, bronzes, precious metals alloys, inks, stones, stamps and more can be studied using a field portable EDXRF system. This manuscript highlights some drawbacks that have to be kept in mind to fulfill a valid measurement such as the need for other backup methods to support portable XRF results or the problem of the non-homogeneity of the sample. This manuscript will also present three examples to demonstrate the usefulness of FP-EDXRF with paintings on canvas, ancient bronzes and sulfur surface analyses.

1. Introduction

In the last two decades the use of mobile or Field Portable (FP) spectrometers, mostly for Energy Dispersive X-Ray Fluorescence (EDXRF), in the archaeometric study of ancient artifacts has become a very common practice as testified by the very large number of papers that appeared in the scientific literature and in the presentations of the main archaoemetric conferences ([1], [2], [3], [4], [5]).
The use of a EDXRF mobile system for the characterization of an ancient artifacts is commonly judged as a precious tool in order to formulate initial hypotheses, to be verified through other kinds of analysis, often destructive. Most often, due to the uniqueness of the object no destructive investigations must be fulfilled and so FP-EDXRF remains the only possibility of gaining some scientific information regarding the elemental composition of the object. As a matter of fact the most attractive thing in this context is the possibility to use the spectrometer in a Non Destructive (ND) manner in order to fulfill the need, often not ineluctable, not to damage the artifact. ND applications of a FP-EDXRF system does open methodological questions about the measurement, which sometimes are not being probed into with the necessary attention and criticism, especially now that often more systems are being used not by scientists.
The errors one might make, while performing a measurement with an EDXRF spectrometer, can be classified in two large groups: conceptual (or methodological) and experimental.
The experimental errors are related to (i) the measuring system and (ii) the procedure adopted during the measurement [6].

Many additional questions arise when the portable instrumentation is used with diagnostic purposes, which require different approaches to the measurement itself.

A diagnostic procedure is the identification of the nature or the cause of a particular process, which can take place through the detection of one of the components of the examined sample, but which also implies a preliminary knowledge of the process. The discrimination is extremely subtle: a diagnostic test assume a prior knowledge of what one wants to observe and thus makes it necessary to explicit the diagnostic query, not to perform a useless measurement. In the case the first hypothesis is not being verified, the obtained results could produce another diagnostic hypothesis. Anyway, this only happens if there is a good knowledge of the process under identification; the practice of "blind" diagnostic tests is nevertheless very common. In other words the diagnostic approach does not increase the knowledge of the process but it helps verifying if it occurs. An important aspect is that the measurement should be systematic and reproducible; in this case the difference with a scientific investigation tends to diminish.

A non-destructive measurement is almost always repeatable but its reproducibility is limited because the exact repositioning is difficult and the local non homogeneity can change the result of the measurement. The study of an ancient artifact should take into account all of these concepts, since the aim of experimental archeometric investigation is frequently not the same as in pure research and in many different applied fields.

The practice in the use of non destructive spectroscopic systems in the archaeometric field leads to recognize the risk to assign incorrect meanings to the results obtained through ND investigations. For example, the homogeneity of the sample is often an implicit assumption in most common analytical techniques; this cannot be true for non-destructive analyses because materials, especially ancient ones, frequently show an internal structure due to both production and degradation processes, and this structure is totally unknown to the scientist. This sets a limit for the precision obtainable while performing measurements on ancient artifacts, unless it is possible to find zones of the object which are meaningful with respect to the investigation and in which the material is homogeneous.

A good knowledge of the material's internal structure and of the degradation processes undergoing, allows to improve the methodology and evaluate the degree of precision of the results that can be obtained. The use of micro-analytical techniques, such as stratigraphy, is most of the times advantageous, but remains meaningful only for the small portion of the abject sampled.

2. Experimental Setup

A FP-EDXRF system is composed of an X-ray source, an X-ray detector and a pulse height analyzer and naturally a computer. A portable EDXRF system must be of small size and light-weight ([7], [8]). For this reason a small size X-ray tube may be employed, coupled with a Si-PIN or a Si-drift X-ray detector. Considering the elements to be analyzed and their concentration in typical works of art, the following systems are currently suggested:

- a small size and low power X-ray tube, working at about 50 kV maximum voltage, and about 100 μA of maximum current; the X-ray tube anode should be selected according to the specific problem;

- a Si-drift detector, with a thickness of about 400–500 μm, and an energy resolution of about 130–150 eV at 5.9 keV or less (the less the better);
- a small size pulse height analyzer.

Filtering the X-ray tube output should be done according to the elements to be analyzed, this operation is very important, with the right filtering it is possible to enhance the signal from needed chemical elements.
When elements emitting low-energy photons (typically from 2 keV to about 5 keV) are analyzed, filtering should be avoided. On the contrary, when elements emitting X-rays from about 5 keV to about 35 keV are analyzed, filtering is needed, and sometimes mandatory, to eliminate the low-energy tail of the incident X-ray spectrum, which only contributes to the background and slows down the system.
Collimation of both X-ray tube and detector is in any case useful; the X-ray tube collimation should be carried out according to the area and concentration of the element to be analyzed. The detector should be weakly collimated, only to avoid scattered photons in the detector.

3. Results

In this report, we show three applications of FP-EDXRF to illustrate the possibilities of the method. The three examples were carried out with non-destructive procedure and *in situ*. The cases are related to paintings, bronzes and stone like materials.
In most of the cases it is evident that other non-invasive techniques are needed to backup FP-EDXRF in order to optimize and enhance its results.

"Peter Paul Rubens's painting "Saint Gregory the Great surrounded by the Saints Papianus, Maurus, Flavia Domitilla, Nereus, and Achilleus" in the Musée de Grenoble"

Non-invasive analyses on the huge canvas of Peter-Paul Rubens "Saint Gregory the Great surrounded by the Saints Papianus, Maurus, Flavia Domitilla, Nereus, and Achilleus" (2.88 m x 4.77 m), which is today in the Musée de Grenoble, France were conducted. The painting was studied for the first time by art historians and scientists together.
In September 1606 Rubens has received from the Oratorian Fathers the commission to paint the high altarpiece for Santa Maria in Vallicella (known also as Chiesa Nuova), their main church in Rome. The commission was linked on a decision to move the Madonna della Vallicella, a venerated, miracle-working fresco dating from the fourteenth century, from a side chapel to the high altar where both images should have appeared together. On 30 January 1608, the painting was refused and Rubens was given the opportunity to complete a new, three-parted ensemble for the altar, finished in October 1608 and still *in situ* (see Figure 1).

Rubens's high altar paintings have received ample attention from art historians for the abundance of surviving archival and pictorial material but no global scientific investigations of Rubens's Grenoble painting were carried out up to now. Imaging and spectrometric techniques have been applied on the huge painting: grazing light, ultraviolet photography, IR reflectography, digital X ray Radiography, EDXRF, SEM/EDS and micro FTIR.

Figure 1: Rubens canvas in Grenoble

The work suffered severe problems due to the dimensions of the painting and due to the fact that the museum room could not be closed for too many days. The canvas was therefore dismantled from the frame and set vertically on the floor along the widest dimension. The first half of the painting was at this point easily analyzed. The painting was rotated and the second half was finally analyzed. The work was done mainly during night because it was impossible during day time to cut the parasite lights. The work in its all took three full nights. To cover the full painting more than 400 UV and infrared photo were made and 166 X-ray radiography were shot (see Figure 2). 32 EDXRF were fulfilled to determine the palette used by Rubens. Availing of the IR and the X ray radiographic results it was possible to sketch the different versions that the painting undergone in the course of its life. It was possible to connect all the versions found in the underlying levels to drawings of Rubens itself or to copies that were made in time.

Figure 2: details of Saint Domitilla in visible light and X ray

For non invasive chemical analyses FP-EDXRF was the only technique available: non invasive, fast, reliable and with immediate results. The UV, IR and RX results were used to successively choose the points to analyze. Along the centuries the painting had received damages that were restored in time. The imaging techniques previously fulfilled in respect to the XRF were necessary to identify points that belonged to the restored sections and points that were a stratification of different versions of the painting. Without this information, the interpretation of the XRF spectra would have lead to wrong results.

In Figure 3 a spectrum of a violet pigment is presented, the complexity of the mixture is evident, a part the white lead present in the preparation, the tone in composed of white lead, red cinnabar, ochre and copper blue while zirconium is a component of the X ray tube collimator.

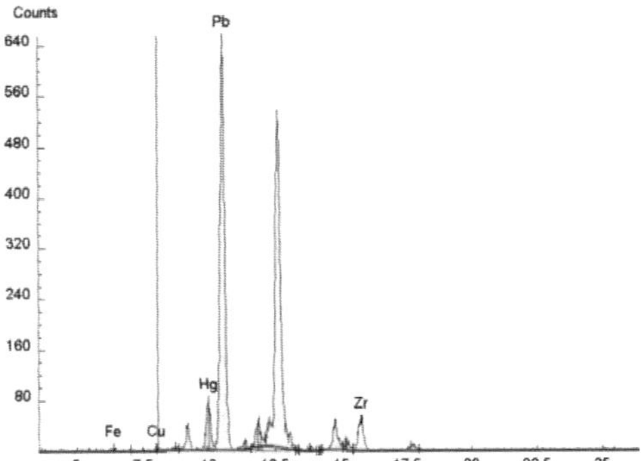

Figure 3: XRF spectrum of a violet pigment in Rubens painting.

The use of XRF technique to define the components of the alloy and patina in a Chinese bronze finishing, in the form of a Taotie Mask (X-VIII Century B.C.E.)

A bronze finishing in the form of a Taotie mask was examined within the ambit of a campaign for the diagnostic investigation and conservation of East Asian works of art.

IOP Conf. Series: Materials Science and Engineering **37** (2012) 012001 doi:10.1088/1757-899X/37/1/012001

The radiographic and subsequent conservation work revealed that the artifact consists of a number of tin/lead-soldered fragments; to clarify possible pertinence and homogeneousness, qualitative-quantitative energy dispersion X-ray fluorescent analyses were carried out.
The analyses revealed that the object is composed of 13 fragments of various compositions; this implies that some parts of the object are the result of restoration integrations. The EDXRF and Principal Component Analysis (PCA) examinations permitted the recognition of fragments in ternary alloy, homogeneous among themselves and, probably, the originals. These are different from the other fragments that present heterogeneous compositions and may, therefore, be considered extraneous. In Figure 4 the thirteen fragments are sketched.

Figure 4: in red EDXRF analyses, in false colors, the subsequent sub-division based on the multivariate data interpretation of the results.

In table 1 the semi quantitative results are presented and in Figure 5 the PCA multivariate results are shown.

Table 1: Semi-quantitative results (peak areas) of the analyzed points.

Measure	Fe	Cu	Zn	Ag	Sn	Sb	Pb
1	607	274240	4840	185	417	566	3971
2	525	182703	3044	128	886	469	1771
3	235	6711	1002	-	12198	78	27298
4	468	191024	2843	131	302	470	1216
5	47	20904	-	-	1912	-	4351
6	268	115217	1794	61	139	216	668
7	352	36754	-	41	3287	-	1453
8	50	43230	-	50	5637	-	1466
10	135	82290	-	-	218	-	477
11	190	140551	-	178	12112	-	4455
12	138	108492	-	118	9475	-	3786
13	113	19370	-	-	1447	-	3310
14	-	6400	-	-	3455	97	9502
15	701	252342	4184	174	172	497	1964
16	267	73870	-	93	7487	-	2622
17	825	210186	3535	167	286	471	3698
18	213	107609	1680	84	461	232	981

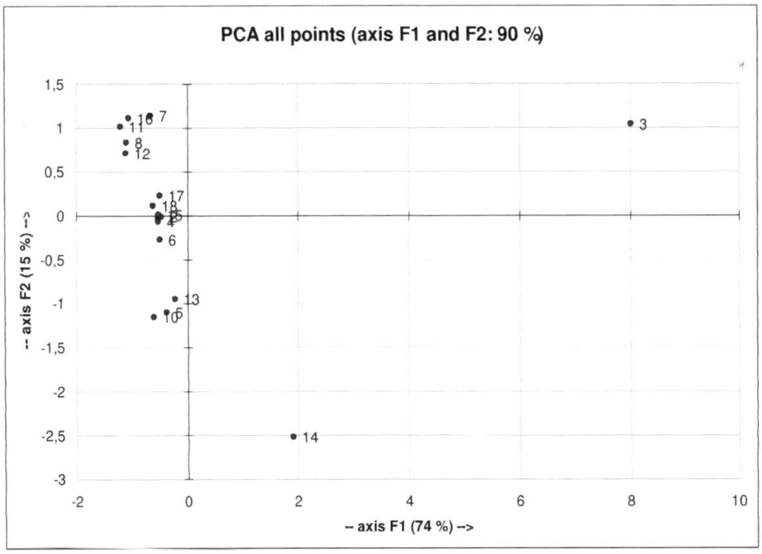

Figure 5: PCA multivariate results on data in Table 1 with outliers.

The graph reveals a net separation between clusters of points (and therefore between fragments). Two outliers (points 3 and 14) that belong to the tin and lead solder and three groups as shown in colors in Figure 4. With the help of an archeologist the group of fragments reported in brown in Figure 4 ended to be original of the Western Zhou period, while the group in dark green appeared to be an ancient restoration. The third group composed by the fragments 5, 10 and 13 may be considered spurious as far as composition and technical-stylistic aspects are concerned.

FP-EDXRF Sulfur-concentration analyses carried out on the facade of the Scrovegni Chapel in Padua.

Within the planned works of preservation, which are previous to a restoration stage, sulfur-concentration analyses have been carried out on the facade of the Scrovegni Chapel in Padua with classical methods and availing of portable non invasive EDXRF system (see Figure 6). The aim of the measurements was to quantify the level of sulfation of the mortars and bricks belonging to the facade. A very dangerous weathering process which deteriorates stone monuments, and is caused by human activity, is air pollution. One of the worst pollutants for carbonated monuments is associated with sulfur compounds and with gypsum. Gypsum is a rather soluble compound that may be washed away by rain. It leaves the surface clean but eroded and open to new corrosive processes. In 1993 the mortars between the bricks that were presenting high sulfur levels were eliminated and at the same time a waterproof protective coating was laid on the facade.

Forty EDXRF measurements were fulfilled, the first 25 in relation to a grid that the restorers had prepared for the monitoring processes among the years. The last 15 spots were chosen *in situ*, during the last measuring campaign. The analyzed spot is 1 square mm and the deepness reached is of few microns, in relation to the low excitation energy of the sulfur atoms. The measurements, totally non invasive, were fulfilled in a quantitative manner availing of portable EDXRF system optimized for the sulfur analyses and a set of sulfur standards. The measurements were fulfilled, furthermore, with the classical methods.

The EDXRF results showed the presence of a gradient with the sulfur concentration over all the facade. The gradient had is maximum at the ground level of the facade and diminished with the height (see Figure 7). Two possible reasons of such a behavior could be connected to the high molecular weight of the sulfur based molecules in respect to the air and to the vertical air streams gradient.

Figure 6: EDXRF sulfur analyses on the Scrovegni facade

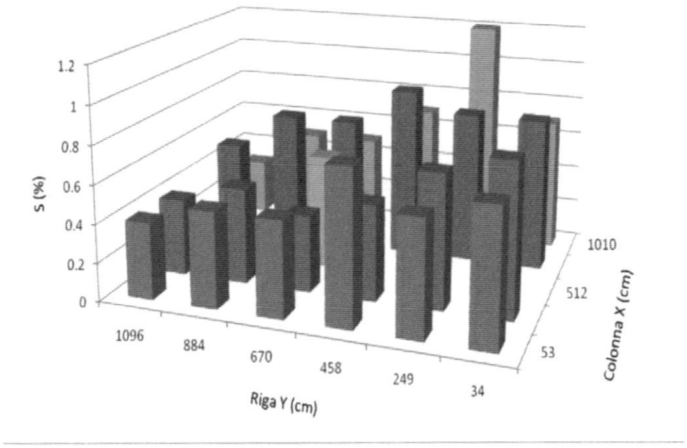

Figure 7: 2D representation of the sulfur results

4. Conclusions

In this manuscript the main drawbacks of FP-EDXRF such as the need for other backup methods to support portable XRF results and the problem of non-homogeneity of samples when non invasive measurements are fulfilled. Three case studies on the use of Field Portable EDXRF are reported on a painting of Peter Paul Rubens, an ancient Chinese bronze mask and the façade of the Scrovegni Chapel in Padua.

5. References

[1] Cesareo R, Brunetti A, Ridolfi S 2008 Pigment layers and precious metal sheets by energy-dispersive x-ray fluorescence analysis *XRS* **37** Issue 4.

[2] Diana M, Gabrielli N, Ridolfi S 2007 Sulfur determination on stone monuments with a transportable EDXRF system *XRS* **36** Issue 6.

[3] Cesareo R, Ridolfi S, Castellano A, Marabelli M, Buccolieri G, Quarta S, Gigante G E 2006. From Giotto to De Chirico to Verrocchio: analyses of paintings and historical bronze alloys availing of portable EDXRF equipment *JNR*, **14** No.1.

[4] Bottaini C E, Silva A L M, Covita D S, Moutinho L M, Veloso J F C A 2012 Energy dispersive X-ray fluorescence analysis of archeological metal artifacts from the Final Bronze Age *XRS* **41** Issue 3.

[5] Ikeoka R A , Appoloni C R, Parreira P S, Lopes F, Bandeira A M 2012 PXRF and multivariate statistics analysis of pre-colonial pottery from northeast of Brazil *XRS* **41** Issue 1.

[6] Gigante G E, Ridolfi S and Ricciardi P 2005 Quantitative analyses of ancient metal artifact by means of portable Energy dispersive x ray fluorescence spectrometers: a critical review *Cultural Heritage Conservation and Environmental Impact Assessment by Non-Destructive Testing and Micro-Analysis* – Van Grieken & Janssens (eds) Taylor & Francis Group, London, ISBN 90 5809 681 5.

[7] Cesareo R and others 2008 Portable systems for EDXRF Analysis of Works of Art *Portable X-ray Fluorescence Spectrometry* The Royal Society of Chemistry ISBN 978-0-85404-552-5.

[8] Roldán C and others 2010 Application of field portable EDXRF spectrometry to analysis of pigments of Levantine rock art *XRS* **39** Issue 3.

Atmospheric composition and micro-climate in the Alhambra monument, Granada (Spain), in the context of preventive conservation

B Horemans[a,1], O Schalm[a], K De Wael[a], C Cardell[c] and R Van Grieken[a]

[a]University of Antwerp, Universiteitsplein 1, BE-2610 Antwerp, Belgium; [b]

Email: benjamin.horemans@ua.ac.be

Abstract. The world famous Alhambra monument in Granada, Southern Spain, listed as UNESCO world cultural heritage since 1984, represents probably the most beautiful example of Islamic art and architecture from the Middle Ages in Europe. It is visited by ca. 2 million people annually. Granada is situated in a natural basin, surrounded by mountains with altitudes up to 3500 m. Due to this topography and the prevailing low wind speeds, pollution-derived and especially traffic-derived particulate matter often accumulates in the urban air. In order to evaluate the potential conservation risks from the surrounding air, the atmospheric composition in the Alhambra monument was evaluated. Indoor temperature and relative humidity fluctuations were evaluated for their potential degenerative effects. Furthermore, the atmospheric composition in the Alhambra was analyzed in terms of inorganic gases (NO_2, SO_2, O_3, and NH_3) and black carbon. It was found that the open architecture protected the indoor environments from developing a potentially harmful microclimate, such as the build-up of humidity resulting from the huge number of daily tourists. On the downside, the strong ventilation made the indoor air hardly different from outdoor air, as characterized by strong diurnal temperature and relative humidity gradients and high traffic-derived pollutant levels.

1. Introduction

The consequences of atmospheric pollution are some of the biggest challenges in modern times. Historical pollution episodes such as the Meuse valley fog (Belgium) in 1930 and the London fog in 1952 made people aware of the causal relationship between air pollution and increased morbidity and mortality [1, 2]. The negative effects of atmospheric pollution do not stop with our health, but also affect the natural balance of many remote areas [3-5], and are directly responsible for climate change phenomena [6-8]. Besides these environmental and health issues, air pollution also accelerates the attack of our tangible cultural heritage.

It is important to know how atmospheric pollution deposits on heritage structures and how it will affect material properties such as morphology, composition, strength, and aesthetic appearance. Atmospheric gases such as SO_2, NO_2 and O_3 are often studied in relation to preventive conservation due to their corrosive effects on many artwork materials [9-11]. Deposition of particulate pollution, also known as soiling, leads to blackening of heritage materials. Blackening is caused by the formation of a thin layer of atmospheric particles which absorb visible light on the surface of diverse construction and decorative materials. Previous publications already reported in detail on the chemical

[1] To whom any correspondence should be addressed.

composition of particulate pollution inside the Alhambra monument [12, 13]. It was found that black carbon (BC) was one of the dominant components of airborne particles with a diameter below 1 μm. Although many atmospheric particles are able to absorb in the visible light spectrum, BC is by far the main component which induces blackening. In addition, soiling of monuments also leads to direct deterioration of heritage materials, due to their interaction with hygroscopic salts [14-17], which also represent a large fraction of the aerosol in the Granada city and the local surroundings of the Alhambra monument [13].

It is evident that the atmospheric degradation of our cultural heritage is extremely complex, depending on many atmospheric conditions and material properties. Nevertheless, it is well known that the atmospheric properties such as climate and composition have a considerable impact on the heritage environment. Stable atmospheric conditions and low pollutant levels are most favorable for the conservation of our cultural heritage. However, in real museum and monument environments, one often encounters conditions which are far from ideal. At the Alhambra, the effects of the semi-open architecture and the huge number of daily tourists on the atmospheric composition, temperature and relative humidity (RH) inside the palaces are considered to be an important threat. This study is a first attempt to evaluate these deterioration risks at the monumental complex of the Alhambra and Generalife in Granada, by analyzing the micro-climate and the pollution load of the local atmosphere. The results could help conservators with formulating targeted strategies with respect to atmospheric degradation.

2. Experimental

Detailed information on the monument, sampling locations, instrumentation and analytical performance data for the present study were outlined in earlier publications and are not repeated here. Only a brief summary of the methodology is given here. Readers looking for more detail are referred to the literature [12].

The Alhambra was a fortified palatine city constructed from the 11th to the 15th century, comprising a fortress, 23 towers, gates, mosques, spas and 7 palaces. The palaces in the monument were built during the Nasrid dynasty (1238–1492). The Nasrid palaces are composed of a complex combination of rooms. Apart from the Royal Baths which are closed from the outdoor environment, all rooms are in direct contact with the outdoor climate. Pollutants, temperature and relative humidity might affect the elaborate polychromed stucco (gypsum-based decoration), marble, wooden ceilings and tiling ornaments inside the rooms. For the semi-open rooms, decorations and volumes of the rooms can be significantly different. Therefore, three rooms (from diverse palaces) and one outdoor reference location (O) have been selected in order to evaluate the impact of the pollution and climatic conditions on the complete complex. Inside the Nasrid palaces of the Alhambra, the 'Room of the Two Sisters' (P1), the 'Royal Baths' (P2) and the 'Ambassadors Hall' (P3) were selected as study areas. Atmospheric pollutants at each of these locations were investigated on a weekly basis, for a total period of three weeks (15th of June until 5th of July 2009).

Every week, parallel samples of major inorganic atmospheric gases (NH_3, SO_2, NO_2, and O_3) were collected at all locations simultaneously, by means of passive sampling devices. The mean atmospheric concentration of the gases was calculated from the sampled mass, sampling time and temperature-corrected sampling rate, as described by the manufacturer of the sampling devices (Radiello, Padova, Italy). Ion chromatography (for NH_3, NO_2 and SO_2) and visible light spectrometry (for O_3) were used for analyzing the absorbed mass of each gas. The concentration of BC was monitored in a semi-continuous fashion (10 min interval) with a portable aethalometer (AE42, Magee Scientific, Berkeley, CA, USA). The instrument was moved each week to a different location inside the palaces (P1, P2, and P3).

The micro-climate of each room was analyzed with calibrated ATX-11 data loggers (ATAL, Purmerend, The Netherlands) which registered thermo-hygrometric parameters every 15 min. These instruments measure the air temperature and RH with a built-in platinum thermo-resistance sensor (Pt1000, accuracy: 0.4°C) and a thin-film capacitive sensor (accuracy: 2.5% in the range of 5-95%

RH), respectively. For comparison, an outdoor reference location was selected to be monitored simultaneously with the indoor locations during a full year.

3. Results

3.1. Meteorological fluctuations

Meteorological conditions affect buildings in several ways. Not only do they influence the rate with which pollutants react with the surface of construction/decorative materials of buildings and the biological activity of micro-organisms, but they also induce material stress through processes of dilatation and contraction. In our study, we expected that the huge numbers of daily tourists visiting the Nasrid palaces of the Alhambra would affect the micro-climate of the diverse rooms, especially the humidity inside the palaces. Moreover, the semi- open architecture of this monument barely buffers the strong outdoor meteorological fluctuations, thereby exposing the unique polychromes and other artwork such as wooden ceilings and tiling to (harsh) outdoor conditions.

The collected time series samples from the three rooms and the outdoor reference locations were very similar. Therefore, the potential danger of the meteorological conditions was analyzed by evaluating the micro-climate at the reference location. The yearly average temperature and RH were found to be 17 ± 8°C (from -1.2 to 33.1°C) and 54 ± 19% (from 28 to 78%), respectively (figure 1). The results of about one year showed only two days (January 9th and 10th) with freezing temperatures during the night (~8 hours). Moreover, temperature never reached the dew point, so condensation on cold walls did not occur. As for RH, about 15% of the year was registered with values above 75.3% (deliquescence point of NaCl crystals), especially in the period from December to February. Long wet periods lasted about 10 to 60 hours and were interrupted with drier (≤ 75% RH) periods or shorter wet periods. The isolines in figure 1 show how temperature and RH fluctuate throughout the day and the year. The average seasonal contribution was estimated by fitting a sinusoidal function to the monthly averaged data (not shown). It was responsible for a total annual temperature and RH change of 20.2°C and 42%, respectively. The average diurnal fluctuation, as estimated from a sinusoidal fit to the hourly averaged data (not shown), showed a total daily change of 4°C and 7%, respectively.

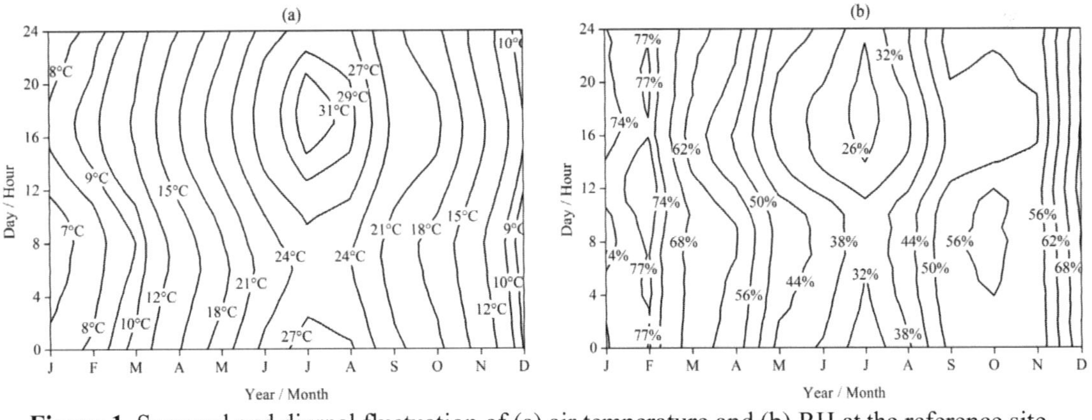

Figure 1. Seasonal and diurnal fluctuation of (a) air temperature and (b) RH at the reference site.

IOP Conf. Series: Materials Science and Engineering **37** (2012) 012002 doi:10.1088/1757-899X/37/1/012002

3.2. Gaseous atmospheric composition

Summer in Granada is characterized by warm and dry days with intense solar radiation and almost no cloud cover. In the first week of the survey, the O_3 concentrations reached the European alarm threshold of 114 ppb, while, during the second and the third week, O_3 levels were below the European warning threshold of 85 ppb (figure 2a). This reduction resulted from a developing low pressure system over the Atlantic Ocean, which disturbed the stable high pressure field maintained over the European continent during the first week. This made the ruling Mediterranean air transport to be gradually replaced by Atlantic air parcels coming from high altitudes (figure 3), which injected fresh air into the city. This change in atmospheric conditions drastically reduced the O_3 formation and accumulation of photochemical smog. Despite this meteorological change, NO_2 and NH_3 were almost constant over the campaign (6-14 ppb), reflecting the constant local traffic source. SO_2 concentrations in and around the Nasrid palaces were extremely low (< 0.45 ppb) and didn't exceed the quantification limit of the sampling method (1 ppb). This is probably due to the strong photoactive oxidation of SO_2, resulting in fast gas-to-particle conversion of sulphur pollutants [18].

At the outdoor reference location (O), O_3 concentrations ranged from 61-111 ppb and averaged around 79 ± 23 ppb (figure 2b). Indoors (P1, P2, P3), the reactive O_3 pollutant is quickly removed from air by deposition on walls and ceilings. An indoor environment which is somehow shielded from outdoor air will therefore have low O_3 levels, since the input of new O_3 does not compensate for the loss due to deposition. This could be seen from the O_3 levels at the more isolated P2 location, which were on average almost two times lower compared to outdoor air. Concentrations at P1 were only slightly decreased, while those at P3 were found to be similar compared to ambient air (figure 2b). This illustrates how the semi-open architecture of the palaces makes the interior highly susceptible for exposure to aggressive pollutants such as O_3. Concentrations for NO_2 inside the palaces were slightly less than outdoors, while for NH_3 the opposite was observed.

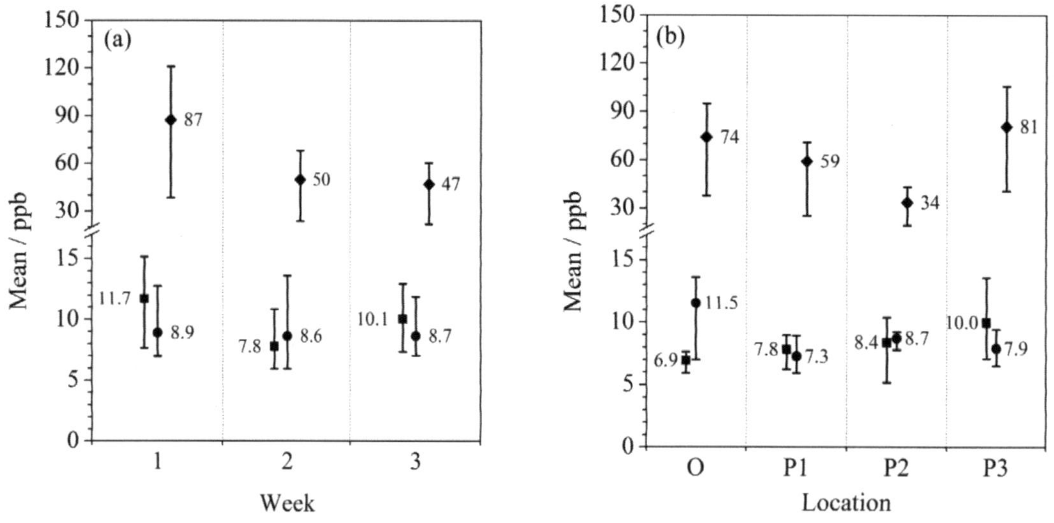

Figure 2. Atmospheric concentration of major inorganic gases averaged according to (a) sampling week (b) location in the monument. Bars represent concentration range. Diamond: O_3; Circle: NO_2; Square: NH_3.

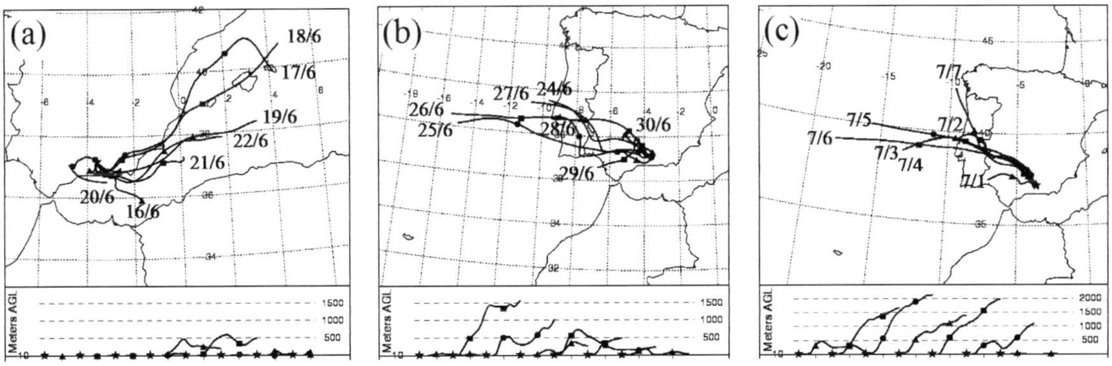

Figure 3. Daily air mass backward trajectories during the (a) first, (b) second, and (c) third week of the sampling campaign, as calculated by the HYSPLIT model with GDAS meteorological data [19, 20]. Trajectories arrive at the Alhambra (37.18° North, 3.59 ° West) on 12 pm UTC of the indicated dates. Below each map, the vertical trajectories are indicated in units of meters above ground level (AGL).

3.3. Black carbon

From the gaseous atmospheric composition, it is clear that the air in the palaces was hardly different from outdoor air. Therefore, it could be expected that the amount of BC also strongly depends on the outdoor conditions. The typical daily pattern of BC inside the palaces is shown in figure 4. It can be seen how the background concentration of BC is fairly constant, around a value of 1 to 1.5 µg m^{-3}. However, at the start of every day (7 to 8 am), BC concentrations increased steeply to a maximal average concentration of 6 µg m^{-3} in a period of only 1-2 hours (maximum around 10 am), after which it dropped slowly. These results illustrate how the Nasrid palaces are directly suffering from exposure to traffic-derived pollutants produced in the local urban area of Granada.

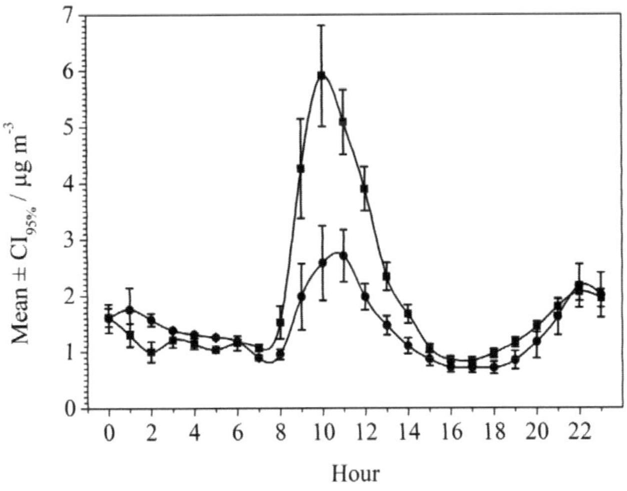

Figure 4. Average daily pattern for black carbon (BC) inside the Nasrid palaces. Bars represent a 95% confidence interval. Square: Monday to Friday; Circle: Saturday to Sunday.

4. Discussion

The collected meteorological time series samples from the three rooms and the outdoor reference locations in the Alhambra monument were very similar. Seasonal and diurnal gradients were present inside the rooms, suggesting that the indoor air is regularly refreshed by outdoor air. Only for the Royal Baths (P2) the impact of outdoor meteorological conditions was rather small, suggesting that the outdoor air diffuses slowly into these rooms. Since the environmental conditions for outdoor and indoor air are so similar, it is difficult to discern the influence of daily tourists on the monument's microclimate, probably due to the fast replenishment of air inside the semi- open rooms, which blows out any humidity excess. However, the outdoor meteorological fluctuations can be considered more aggressive to the wooden and stucco painted decorations as compared to typical conditions in museum environments, which are often equipped with instrumentation for climate control.

The major difficulty of the present study is the semi-open architecture of the monument, and hence, nearly identical indoor and outdoor conditions, while ideal meteorological conditions for the conservation of heritage objects are usually reported for confined indoor environments. It is suggested that mixed objects should be stored at temperatures between 19 and 24°C and at a RH ranging from 45 to 60% [21]. The annual average temperature at the Alhambra was found to be slightly below this ideal indoor temperature range (17°C), while RH seemed to be acceptable (54%). Inside the studied rooms material damage due to freeze-thaw cycles and condensation on cold walls was found to be negligible. Also, damage due to crystallization-dissolution cycles of salts inside porous materials [22] was limited, since the RH throughout the year was generally below 75.3%, i.e. the deliquescence point of NaCl. Only in the period from December to February, the RH exceeded the critical limit of 75% on a regular basis (figure 1).

However, the most disastrous effect of meteorological parameters to objects of art in the Alhambra is the rate with which they fluctuate. The fast and unpredictable changes in temperature and RH was decomposed in a factor describing seasonal, diurnal and weather effects. While the former two are periodic fluctuations depending on solar radiation, the latter is the result of (rapid) changes in meteorological conditions.

In general, linear temperature and RH gradients not exceeding 1°C and 5-10% per day, respectively, are considered as acceptable. Seasonal gradients occur gently and were found to be maximally 0.18°C and 0.34%, respectively. Diurnal cycles were characterized by strong temperature and RH changes due to the particular topography of the Granada area, which is situated in a natural basin surrounded by mountains with altitudes up to 3500 m. Inside the Nasrid palaces, maximum temperature gradients in the period from November until March ranged from 1.7 to 3.1°C day^{-1}, while RH was almost constant throughout the day (figure 1). In spring until late summer (April until October), maximum diurnal temperature gradients ranged from 4.8 to 6.1°C day^{-1}, while RH gradients were 9.4–12.8% day^{-1}. Especially for materials with a short response time (hours), such as paint layers, such changes might be harmful since they are sensitive to expansion and shrinkage of the surface material [23]. For materials with a longer response time (few days), such as wooden objects with more than 1 cm thickness, high frequency changes occur probably too fast for the object to adapt to the new conditions. The most dangerous phenomenon is probably the change in weather conditions. They are slow enough, so materials can adapt themselves to the environment, but they are fast enough to cause material stress, and hence, structural damage [24].

During the analysis campaigns, there was considerable air flow through the semi-open rooms, which blows out any humidity and gases produced by the visitors. However, the strong ventilation also results in exposure to pollutant levels which are typically found at outdoor environments. The composition of typical museum atmospheres is generally found to be poor in gaseous pollutants, with levels for O_3, NO_2 and SO_2 below 10 ppb, between 30 and 70 ppb and less than a few ppb, respectively [25]. At the Alhambra monument, O_3 levels were high inside the Nasrid palaces, especially during the first week of the survey. It is well known that O_3 reacts with organic painting materials such as dyes, binders or surface coats, leading to the fading of color paint layers [26]. Since 1923 until recently, all polychromed carpentry in the Alhambra was treated with a mixture of linseed oil, turpentine and beeswax to protect it against deterioration [27]. Ironically, in a way, this organic mixture could have enhanced the fading of the colors due to reaction with O_3. For the inorganic pigments of the polychrome stucco and carpentry artworks of the Alhambra, such as vermilion, red lead, malachite and azurite [27, 28], the effects of O_3 are still unknown. Concentrations of NO_2 and SO_2 in the palaces were similar to those found at other museum environments [29]. These acidifying gases are found to corrode artwork upon deposition.

On the other hand, soiling of buildings is one of the major problems related to particulate pollution. Deposition of particles affects the aesthetic appearance of artwork, by absorbing incident light. BC is by far the strongest absorber of visible light of all the components in airborne particles. Reducing the deposition of these black particles on monuments is therefore a priority issue for many conservators. In the Alhambra palaces, indoor BC concentrations depended strongly on traffic conditions in the Granada area (figure 4). On average, the background concentration of BC was found to be around 1 $\mu g\ m^{-3}$. When considering an average indoor deposition rate of 0.03 $cm\ s^{-1}$ for particles below 2.5 μm [30], the background concentration accounts for a total soiling of ~0.95 mg BC $m^{-2}\ year^{-1}$. However, daily concentrations were raised on average by a factor of 3.5 above this background for a total period of 8 hours per day. Such an increase could enhance soiling by about 80%, raising it up to 1.7 mg BC $m^{-2}\ year^{-1}$. Since these calculations are based on the average BC concentration during three weeks, it must be stressed that the extrapolated deposition rates of one year should be treated with care. Nevertheless, it gives a rough quantitative estimate on the soiling due to black particles in the Alhambra, and shows how the blackening process could be diminished by simply reducing the BC emissions close to the monument.

5. Conclusion

The evaluation of atmospheric risks for indoor environments is usually made by comparing quantities such as average temperature, relative humidity and maximum concentrations of pollutants with threshold values. In the case of semi-open rooms exposed to the outdoor climate, such evaluation is much more complicated. It was anticipated that the huge number of tourists visiting the Alhambra monument could affect the climate in the Nasrid palaces. Nevertheless, the strong ventilation and open architecture of the palaces made temperature and RH hardly different from outdoor air, and prevented the build-up of high humidity, condensation and microbial activity. However, at the downside, indoor decorations of the monument are exposed to strong meteorological fluctuations. While seasonal variations are still acceptable, strong diurnal temperature gradients are believed to pose a risk to the conservation of the polychrome and wooden artwork all over the monument. Analogously, the atmospheric composition inside the palaces was almost identical to that of the outdoor air. This was especially true for O_3, which was found with high (smog-like) concentrations. BC was also found at higher concentrations than one could tolerate in a cultural environment.

Recently, Granada City Hall is planning to help tourists to get more easily to the monument by allowing cars to drive through the Alhambra Park, crossing the 'Gate of the Pomegranates', located on a steep street. At present, the BC concentration close to the entrance of the park is about 2 $\mu g\ m^{-3}$, and about 1 $\mu g\ m^{-3}$ inside the Nasrid palaces of the Alhambra. However, if the street is opened for traffic in the future, the BC concentration is expected to rise up to 8 $\mu g\ m^{-3}$ or even higher, as found at a similarly steep street with dense traffic in the nearby Albayzín quarter (world heritage site since 1984). Such decisions could have a considerable impact on the levels of BC and other vehicle derived pollutants such as NO_2, SO_2 and O_3 inside the Nasrid palaces, with major implications for their future conservation and the enjoyment of visitors to this famous monument. By simply limiting the traffic in and around the monumental complex of the Alhambra and Generalife, soiling and blackening of this unique Muslim monument could be minimized, thereby cutting on the budget for conservation and restoration of the highly detailed polychromed wall and ceiling decorations.

Acknowledgement
The authors thank the research group RNM-179 (CICE, JA, Spain) and the 'Patronato de la Alhambra y Generalife' for making this research possible. Special thanks go to M. Urosevic and J. Romero-Pastor for their help in data acquisition. The authors gratefully acknowledge the NOAA Air Resources Laboratory (ARL) for the provision of the HYSPLIT transport and dispersion model and the READY website (http://www.arl.noaa.gov/ready. html) used in this publication.

References

[1] Nemery B, Hoet P H M and Nemmar A 2001 *Lancet* **357** 704-8
[2] Logan W P D 1953 *The Lancet* **261** 336-8
[3] Bencs L, Krata A, Horemans B, Buczynska A J, Dirtu A C, Godoi A F L, Godoi R H M, Potgieter-Vermaak S and Van Grieken R 2009 *Atmos. Environ.* **43** 3786-98
[4] Bargagli R 2008 *Sci. Total Environ.* **400** 212-26
[5] Fromberg A, Cleemann M and Carlsen L 1999 *Chemosphere* **38** 3075-93
[6] Ramanathan V and Feng Y 2009 *Atmos. Environ.* **43** 37-50
[7] Lee D S, Fahey D W, Forster P M, Newton P J, Wit R C N, Lim L L, Owen B and Sausen R 2009 *Atmos. Environ.* **43** 3520-37
[8] Kondratyev K Y 1996 *Energy Convers. Manage.* **37** 763-8
[9] Kontozova-Deutsch V, Godoi R H M, Worobiec A, Spolnik Z, Krata A, Deutsch F and Van Grieken R 2008 *Microchim. Acta* **162** 425-32
[10] Gysels K, Delalieux F, Deutsch F, Van Grieken R, Camuffo D, Bernardi A, Sturaro G, Busse H-J r and Wieser M 2004 *J. Cult. Herit.* **4** 221-30
[11] Godoi R H M, Kontozova V and Van Grieken R 2006 *Atmos. Environ.* **40** 1255-65
[12] Horemans B, Cardell C, Bencs L, Kontozova-Deutsch V, De Wael K and Van Grieken R 2011 *Microchem. J.* **99** 429-38
[13] Potgieter S, Horemans B, Anaf W, Cardell C and Van Grieken R 2012 *J. Raman Spectrosc.* In press. DOI 10.1002/jrs.4052
[14] Cardell C, Rivas T, Mosquera M J, Birginie J M, Moropoulou A, Prieto B, Silva B and Van Grieken R 2003 *Earth Surf. Proc. Land.* **28** 1-14
[15] Cardell C, Delalieux F, Roumpopoulos K, Moropoulou A, Auger F and Van Grieken R 2003 *Constr. Build. Mater.* **17** 165-79
[16] Cores. Consumption of oil-based products. 2008 [cited 2011 24 May]; Available from: http://www.cores.es.
[17] Rivas T, Prieto B, Silva B and Birginie J M 2003 *Earth Surf. Proc. Land.* **28** 425-36
[18] Kelly T J, McLaren S E and Kadlecek J A 1989 *Atmos. Environ. (1967)* **23** 1315-32
[19] Draxler R R and Rolph G D. HYSPLIT (HYbrid Single-Particle Langrangian Integrated Trajectory) Model access via NOAA ARL READY Website. Silver Spring, MD: NOAA Air Resources Laboratory; 2003 [cited 2011 6th June]; Available from: http://www.arl.noaa.gov/ready/hysplit4.html.
[20] Rolph G D. Real-time Environmental Applications and Display sYstem (READY) Website (http://www.arl.noaa.gov/ready/hysplit4.html). Silver Spring, MD: NOAA Air Resources Laboratory; 2003.
[21] Alcántara R. Standards in preventive conservation: meanings and applications. Report: ICCROM - International Centre for the Study of the Preservation and Restoration of Cultural Property 2002.
[22] George W S 2004 *Cem. Concr. Res.* **34** 1613-24
[23] Camuffo D. Microclimate for Cultural Heritage. Amsterdam: Elsevier; 1998.
[24] Thomson G. The museum environment. Cornwall: Butterworth-Heinemann; 1986.
[25] Brimblecombe P 1990 *Atmos. Environ. B-Urb.* **24** 1-8
[26] Whitmore P M and Cass G R 1988 *Stud. Conserv.* **33** 29-40
[27] C. Cardell C, Rodriguez-Simon L, Guerra I and Sanchez-Navas A 2009 *Archaeometry* **21** 637-57
[28] Cardell-Fernandez C and Navarrete-Aguilera C 2006 *Stud. Conserv.* **51** 161-76
[29] Brimblecombe P 1990 *Atmospheric Environment. Part B. Urban Atmosphere* **24** 1-8
[30] Lai A C K 2002 *Indoor Air* **12** 211-4

IOP Conf. Series: Materials Science and Engineering **37** (2012) 012003 doi:10.1088/1757-899X/37/1/012003

Data and interpretation: enhancing conservation of art and cultural heritage through collaboration between scientist, conservator, and art historian

Jo-Fan Huang[1]

[1]Paper Conservator, Department of Conservation, Abu Dhabi Tourism and Culture Authority (ADTCA)

E-mail: jofanhuang@gmail.com

Abstract. Conservation practices can be greatly enhanced and influenced by scientific analysis and art historical insights. In the same respect, scientific data can be contextualized and substantiated by findings from visual examination and historical research. Such collaboration can contribute to the field of conservation in multiple ways: by assisting the conservator to investigate treatment options, discover artists' materials and techniques, determine date of manufacture, and investigate conservation treatment materials. Several technical studies conducted by the author and her collaborators employed micro-x-ray fluorescence (μ-XRF), Raman Spectroscopy, Fourier-transform infrared spectrometry (FTIR), and scanning electron microscopy with energy dispersive x-ray spectroscopy (SEM-EDS) and polarized light microscopy (PLM). These techniques were used on the following previously published projects such as a Japanese painting at the Museum of Fine Arts at Boston, Thai manuscripts at the Arthur M. Sackler Museum at Harvard University, and Chinese export paintings on pith at the Winterthur Museum. Although these studies have been published in the past, they are worthy examples to illustrate how collaborations between conservator, scientist, and art historian complement one another. This presentation will also touch upon ethics in sampling of fine art materials and several online databases such as Infrared and Raman User Group (IRUG) and Conservation and Art Materials Encyclopedia Online (CAMEO), which have proven to be very helpful in the field of conservation.

1. Introduction

Analyzing works of art is not always as easy as it appears. Studies of the physical fabric, the manufacture, chemical make-up, and even deterioration process are all inseparable to the artworks' age, previous storage environment, provenance, artist's technique, and previous restoration, etc. Therefore, a holistic approach to research is imperative to one's understanding of the cultural heritage at hand.

In the past few years, various workshops and conferences have placed emphasis on the collaboration between the science and art. In particular, the workshop titled *Chemistry and Materials Research at the Interface between Science and Art*, funded by the Andrew W. Mellon foundation and the National Science foundation, gathered scientists from the US and Europe and provided a road map for the field of conservation science. A report from this workshop lays out three grand challenges in the field of conservation and conservation science [1]: 1) Materials and Structural Characterization of Cultural Heritage Objects, 2) Understanding Material Degradation and Aging, 3) Materials Stabilization, Strengthening, Monitoring, and Repair. Inspired by the workshop, this paper will discuss some of these points and try to demonstrate why a holistic approach towards research can contribute greatly to the field of conservation, science, and art history.

2. Ethics for conservators and conservation scientists regarding scientific analysis

Consulting various professional guidelines from different national and international conservation organizations, the consensus support on minimal sample is solid. In many guidelines, the conservators are encouraged to conduct research and to understand methods of scientific research.

The conservation community in the United States has adapted the general approach of "less is more." This concept, developed since the mid 1970s, has now become a staple in the conservation practices [2]. The American Institute for Conservation (AIC) has imposed a detailed code of ethics and guideline of practice on their members regarding scientific analysis [3]. Any research is instigated by close examination and sufficient justification. Since the physical integrity of the object should be kept intact as much as possible, non-destructive analysis is always considered prior to sampling. Should sampling become unavoidable, with the permission of the curator and owner only a minimal amount of sample is taken from an inconspicuous area. The remainder of the sample should be kept as part of the documentation. It is worth noting that some institutions may impose even more stringent guidelines on sampling based on their internal policy and procedures.

3. Suzuki Harunobu's *Spring Outing on the Banks of the Sumida River* [4]

Suzuki Harunobu (1724-1770) was a prolific artist and one of the first ukiyo-e print artists to employ full-color printing in Japan during the Edo period (1603-1868). However, little information on paintings by Harunobu has been available perhaps due to the relatively few paintings he made and the even fewer that survive today. Harunobu's *Spring Outing on the Banks of the Sumida River* from the Museum of Fine Arts, Boston, underwent conservation treatment in preparation for an exhibition. Because of extensive media abrasion and loss, areas of the painting had been over-painted with white paint that stood out against the overall aged background. Although inpainting was applied in an obvious crude manner contrasting with the artist's typical meticulous style, a study was carried out to determine which areas had been inpainted in order to aid the conservator and curator in making appropriate decisions on conservation treatment. Visual examination with visible, ultraviolet, and infrared light examination was conducted by the conservator. X-ray fluorescence (XRF), polarized light microscopy (PLM), and Fourier-transform infrared spectrometry (FTIR) was conducted by the scientists, Michelle Derrick and Richard Newman, to study the artist's palette.

Under the magnification, red inclusion was found in the crudely inpainted areas. In contrast, the meticulous painted white areas were homogenous. The analysis showed that both types of white contain calcite, and the PLM and XRF confirmed vermilion was used in the white inpaint areas. The Conservation and Art Materials Encyclopedia Online (CAMEO) database from MFA, Boston was used to confirm the PLM result (http://cameo.mfa.org/). In addition to the analysis, visual examination using Infrared Reflectography (IR) revealed Harunobu's detailed under drawing under the inpaint as demonstrated in figure 1 left.

IOP Conf. Series: Materials Science and Engineering **37** (2012) 012003 doi:10.1088/1757-899X/37/1/012003

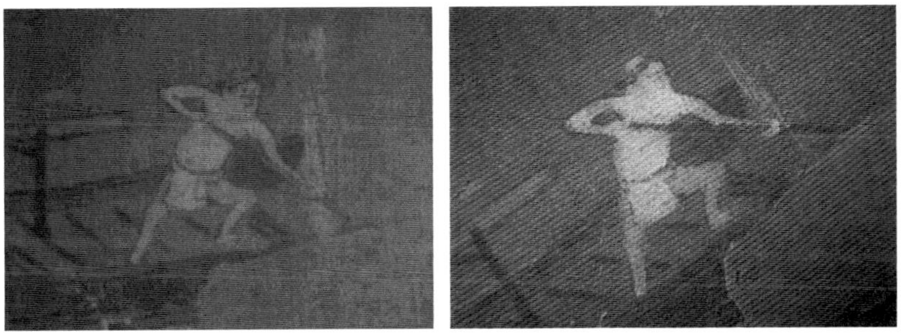

Figure 1: IR photomicrograph revealing Harunobu's fine brush strokes (left); the same view under visible light (right), 40x.

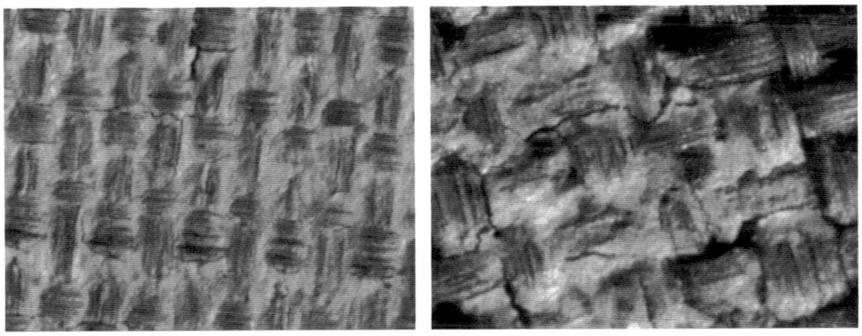

Figure2: Comparison between the original white paint (left) to inpainting (right), 90x

A woodblock print reproduction of the painting was found as an illustration in *Kokka Gakkai Zasshi* (edition 49: 1893), a Japanese art journal in publication since the Meiji period. This print, possibly a direct copy of the painting, served as a point of reference showing the condition of the painting at an earlier time, which has provided helpful visual comparison to identify the areas of restoration.

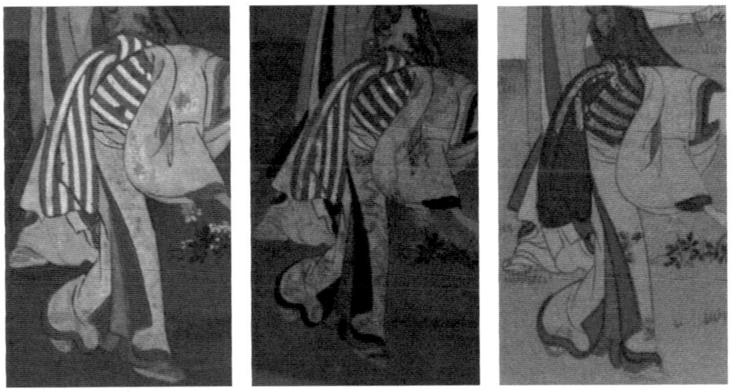

Figure 3: Comparison between the original white paint (left), the same view under the uv light (middle), and a reproduction print. Magnification unknown.

Based on these findings, the conservator was able to document areas of inpaintings. The conservators and the curators have decided to keep the inpainted areas intact to avoid accidental dislodging of the original media.

4. Technical study of four Chinese export paintings on Pith [6]

Four Chinese export paintings on pith (*Tetrapanax papyriferum*) belonging to the Winterthur Museum study collection were examined and analyzed as part of the author's second year graduate course work. Due to the little attention generally spent on this sort of object, the study aimed to expand the knowledge of this unique type of artwork by focusing on artist's techniques and materials as well as overcoming challenges while analyzing this atypical painting substrate.

These export paintings on pith often depicted scenes of Chinese customs and industries, and were popular among the Westerners between the 1790s and 1920s prior to the spread of photography in China. Due to the workshop system, the production of painted pith was a collaborative process rather than an individual effort. Close visual examination has revealed that paints were applied layer by layer with little mixing, as traditionally each painter was assigned motifs or sections to complete with a limited palette. Sometimes the initial sketches were printed by woodblock onto the pith instead of outlining with a brush and ink. Different color paints were applied on both verso and recto of the pith, and a similar technique was also found in Chinese painting on silk.

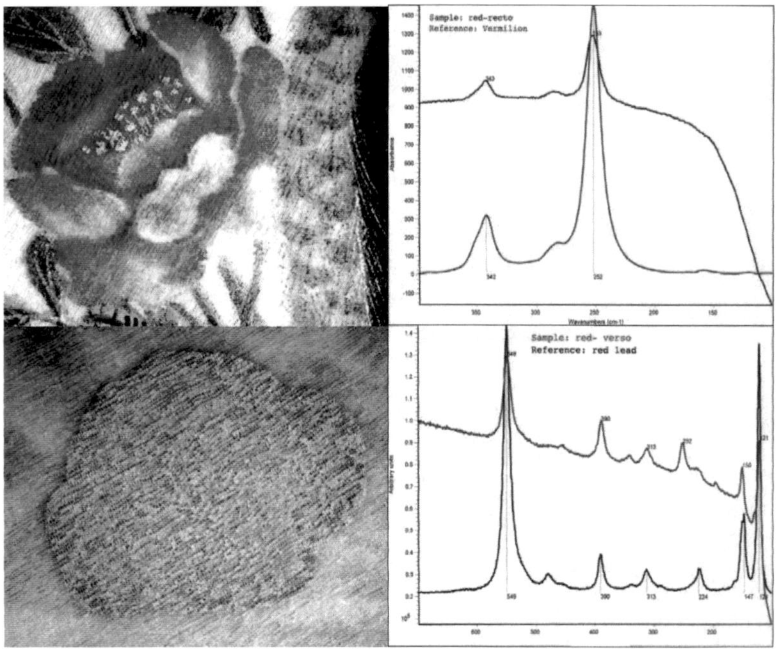

Figure 4: Clockwise from top left: vermilion paint on the front of the pith and its perspective spectra; red lead was found on the verso and its perspective Raman spectra

Pith is a veneer from the core of a plant, with a thickness of 1-2 mm and a porous surface resembling honeycomb. This posed a challenge in non-destructive analysis, for the XRF could not singly detect elements on the recto or the verso of the pith. As the curator did not want any sampling done on one of the paintings, Raman spectroscopy enabled the scientist to conduct surface analysis, 1 to 2 microns penetration, and a small spot size as little as 2 microns. The data was compared to private reference library and public reference libraries such as Infrared and Raman User Group (IRUG). The analysis has indicated that besides lead white, which was occasionally mixed in with colors, the painters' palette was exceptionally homogeneous. The basic palette includes lead white, azurite, malachite, red lead, vermilion, ultramarine, Prussian blue, gamboge, and ivory black. This finding was consistent with art historians' claims on workshop practice and assembly line style production method.

5. Technical study of Thai manuscripts in the 18th, 19th, and 20th centuries [7, 8]

Seven of 55 Thai manuscripts belonging to the Philip Hofer collection of Harvard University Art Museums underwent analysis to determine the date of manufacture. The Thai manuscript collection has received little attention in the past, and only a small number of scholars have studied them. In 2005, several Thai manuscripts from the Hofer collection were loaned for an exhibit, *The Kingdom of Siam: The Art of Central Thailand, 1350-1800*, at the Asian Art Museum of San Francisco in 2005. One particular manuscript, *Illustrated Jataka Tales* (1984.511), was singled out and highlighted in the catalogue. Scholars found one of the images in this manuscript resembled a banner in San Francisco's collection, which had interesting implications for workshop practices. Thus, a technical study on this manuscript and banner was planned at both institutions to investigate and compare the artists' materials and techniques. Unfortunately, further investigation in San Francisco discovered that their banner was a copy of the Hofer collection manuscript made by a Buddhist monk in the 1960s. However, the discussions about the manuscripts highlighted the need for an in-depth technical study of the materials and artists of Thai manuscripts. Based on existing museum record, manuscripts allegedly from the 18th, 19th, and 20th centuries were analyzed using XRF, Raman Spectroscopy, FTIR, SEM-EDS, and PLM. Conservators also conducted visual examination and conduct historical research on artists' techniques and materials.

The analyses indicate the presence of huntite, lead white, calcite, indigo, red lead, vermilion, gamboge, ultramarine, and red earth in the 18th century manuscripts. In the 19th and 20th century manuscripts, the analyses indicate lead white, calcite, red lead, vermilion, gamboge, chrome yellow, emerald green and Prussian blue, as well as possible barium sulfate, titanium white, and viridian. Lead white, gold leaf, and vermilion were found consistently in all seven manuscripts. The results also suggest that different blue and green pigments were used in different centuries: indigo and copper citrate in the 18th century; Prussian blue or synthetic ultramarine were the primary blues in 19th and 20th century; Emerald green began to appear in the artists' palette around mid-19th century; mixtures of Prussian blue or synthetic ultramarine with either gamboge or lead chromate were found in manuscripts dated 19th century.

Table 1. 7 Thai manuscripts with their given and analytical dates

Manuscript accession number	Title	Given date	Analytical date
1984.511	Illustrated Jataka Tales	Late 17th to 18th century	No change
1984.512	Illustrated Manuscript	18th century	No change
1984.508	Illustrated Ten Jataka Tales	Early 19th century	Late 19th century
1984.501	Illustrated Dharma Quotes/ Ten Jataka Tales	Late 18th to early 19th century	No change
1984.517	Illustrated Text on Cosmology	Late 19th century	No change
1984.524	Illustrated Phra Malai Manuscript	19th century	Late 19th to 20th century
1984.510	Illustrated Phra Malai Manuscript	Early 20th century	Mid-19th century?

Two manuscripts were found to be made later than the museum record due to the discovery of emerald green and lead chromate, for several historical records suggested that emerald green and lead chromate were only made commercially available in China and the region in the late 1840s to early 1850s. In addition, the identification of copper citrate was very exciting, since it had never been found in works of art in any of previous research.

6. Conclusion

This paper only touches upon a small section of this very large topic. The possibility for collaboration between the scientific community and conservators is abundant. The case studies have shown that the scientific analysis of works of art can aid the curator and conservator in making appropriate decision on conservation treatment and complement art historical findings to enrich our understanding of artist's methodology and materials, which subsequently answer a question of dating works of art.

Placing the emphasis on the region, as the UAE has a vastly growing artist community, one must also think about modern materials. Scientific analysis of modern materials has gained momentum within the scientific community in the US, UK, and Europe; it also deserves much attention from the scientific community in this region as well. In addition to the analysis, the conservator could collaborate with art historians to conduct artists' interviews and document their techniques and materials, and furthermore their preferences for preservation.

Acknowledgements

I would like to thank my past collaborators and mentors who have made each study possible: Michele Derrick, Richard Newman, Jacki Elgar, Philip Meredith, and Tanya Uyeda from M.F.A. Boston; Dr. Jennifer Mass, Catherine Matsen, Jan Carlson, and Betty Fiske from Winterthur Museum; Dr. Gene Hall from Rutgers University; Dr. Katherine Eremin, Dr. Jens Stenger, and Dr. Narayan Khandekar from Harvard University Art Museums.

References

[1] *Chemistry and Materials Research at the Interface Between Science and Art: Report of a Workshop Cosponsored by the National Science Foundation and the Andrew W. Mellon foundation.* July 6-7, 2009 http://mac.mellon.org/NSF-MellonWorkshop/2009%20NSF%20final%20report%20low%20res.pdf (accessed Nov. 15, 2011)
[2] Stoner J. 2005. *Changing Approaches in Art Conservation: 1925 to the present. Scientific Examination of Art: Modern Techniques in Conservation and Analysis*, National Academic Press. http://www.nap.edu/catalog/11413.html (accessed July 27, 2011)
[3] American Institute for Conservation Code of Ethics and Guideline of Practice http://www.conservation-us.org/index.cfm?fuseaction=page.viewpage&pageid=1026(accessed November 14, 2011)
[4] Huang J. *A Technical Study of a Painting by Suzuki Harunobu at the Museum of Fine Arts, Boston: Artist's Palette and Techniques*. Poster from The Book and Paper Group Annual, Vol 26:2007.
[5] Harvard University Herbaria. http://www.huh.harvard.edu/libraries/Tetrap_exhibit/pith_about.htm (accessed March, 2012)

[6] Huang J., Mass J., Carlson J. et al. *A Technical Study of Chinese Export Paintings on Pith*. 2004. Post-print of North American Graduate Programs in the Conservation of Cultural Property (ANAGPIC).
[7] Eremin K., Jens S, & Huang J, et al. 2008. *Examination of pigments on Thai manuscripts: the first identification of copper citrate*, Journal of Roman Spectroscopy, volume 39, Issue 8, p.1057-1065
[8] Huang J. A Technical Examination of 7 Thai Manuscripts in the 18th, 19th, and 20th Centuries, 2006, Post print of ANAGPIC. www.ischool.utexas.edu/~anagpic/studentpapers2006.htm

Characterization of historic mortars and earthen building materials in Abu Dhabi Emirate, UAE

Benjamin L. Marcus[1]

[1]Building Conservator, Abu Dhabi Authority for Culture and Heritage (ADACH)

Email: benjamin.marcus@adach.ae; blmarcus@gmail.com

Abstract. The Abu Dhabi Authority for Culture and Heritage (ADACH) is responsible for the conservation and management of historic buildings and archaeological sites in the Emirate. Laboratory analysis has been critical for understanding the composition of historic materials and establishing appropriate conservation treatments across a wide variety of building types, ranging from Iron Age earthen archaeological sites to late-Islamic stone buildings. Analysis was carried out on historic sites in Al Ain, Delma Island and Liwa Oasis using techniques such as micro-x-ray fluorescence (MXRF), scanning electron microscopy/energy dispersive x-ray spectroscopy (SEM-EDX), polarized light microscopy (PLM), and x-ray diffraction (XRD). Testing was conducted through consultant laboratories and in collaboration with local universities. The initial aim of the analysis was to understand historic earthen materials and to confirm the suitability of locally sourced clays for the production of mud bricks and plasters. Another important goal was to characterize materials used in historic stone buildings in order to develop repair mortars, renders and grouts.

1. Introduction

Since its establishment in 2005, the Abu Dhabi Authority for Culture and Heritage (ADACH) has been working to protect and conserve Abu Dhabi's built heritage. With over 90 listed historic buildings and hundreds of major and minor archaeological sites spread across a large territory, the scope of ADACH's work is large and continues to grow as new sites are added through survey and excavation. Traditional materials used in the construction of Abu Dhabi's historic buildings include earthen bricks and plasters, stone masonry with lime and gypsum mortars and renders, and wooden elements made from palm and hardwoods such as mangrove.

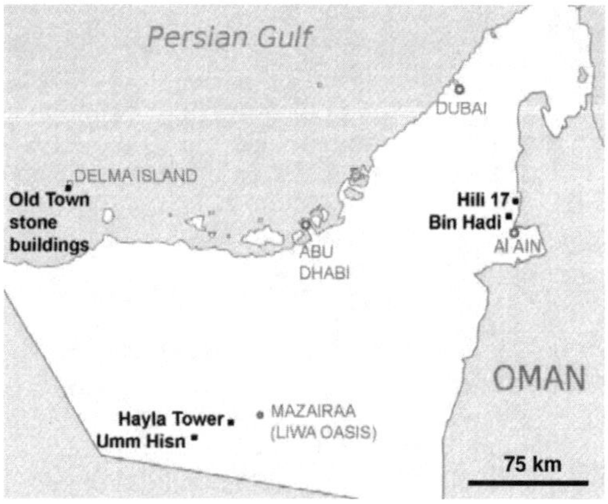

Figure 1: Map of the UAE showing site locations.

While there is a rich oral tradition of construction techniques and materials that has been passed through generations, few scientific studies of historic buildings were made in Abu Dhabi before the establishment of ADACH. Laboratory analysis of architectural materials has complemented the oral tradition and played a key role in understanding historic buildings and developing appropriate conservation treatments. This paper presents scientific analysis of building materials carried out by consultants and ADACH staff over a period of five years beginning in 2007. Each site has been the

Published under licence by IOP Publishing Ltd

focus of comprehensive assessment and planning, though this paper will focus mainly on materials and how analysis results informed conservation. The studies include two stone towers in Liwa Oasis: Hayla Tower and Qalat Umm Hisn; two earthen sites in Al Ain: the Hamad bin Hadi al Darmaki House in Hili Oasis and Hili 17, an Iron Age archaeological site located in Hili Archaeological Park; and a group of stone buildings located on Delma island that includes three mosques and a house museum (see map, Fig. 1).

2. Stone buildings and mortars

2.1. Bait al Muraykhi and the Al Muraykhi, Al Muhannadi, and Al Dawasari mosques, Delma Island

In 2009 ADACH began a conservation management plan (CMP) for four historic stone buildings located in the Old Town of Delma Island: the Bait al Muraykhi, or Pearl Merchants House, and the mosques of Al Muraykhi, Al Dawasari and Al Muhannadi (Fig. 2). Built around 1930, the decorated coral stone and plaster buildings reflect the wealth of the pearl trade, the main economic livelihood of Delma Island in the 19th and early 20th centuries. Pearling on Delma Island declined following World War Two and by the late 1980s the buildings had fallen into disrepair, with only one of the mosques still in active use. In 1993-94 all four buildings were restored and the Pearl Merchants House turned into a museum. However, the local coral stone and gypsum-based mortars used in the original construction and subsequent restoration contain salts that have caused significant damage to the buildings and their foundations. Structural problems, termites and the humid marine environment have also contributed to the deterioration of the exteriors and wooden elements such as doors and roof beams.

Figure 2: Clockwise from top left: Bait al Muraykhi and the Al Muraykhi, Al Muhannadi, and Al Dawasari mosques. Source: Donald Insall Associates (DIA).

CHEMICAL ANALYSIS	RS06	RS08	RS10	RS11
Insoluble Residue	0.25	0.25	4.80	8.45
Soluble Silica, SiO_2	3.69	3.28	3.79	3.94
Acid soluble Alumina, Al_2O_3	0.77	0.67	0.80	0.83
Acid soluble Iron, Fe_2O_3	0.05	0.07	0.09	0.11
Acid soluble Calcium, CaO	28.95	31.84	36.31	32.80
Acid soluble Magnesium, MgO	1.17	1.05	1.72	1.62
Acid soluble Sulphate, SO_3	49.92	40.37	30.20	31.91
Loss on Ignition	15.11	22.37	21.51	20.13
Total	99.91	99.90	99.22	99.79

Table 1: Chemical analysis results of four render samples showing predominance of gypsum ($CaSO_4{\cdot}2H_2O$). Source: DIA.

Following completion of the CMP this year, work began at the site on sourcing and testing mortars and renders for stabilizing the lower walls of the Bait al Muraykhi, which are undermined by rising damp and salts. Conservation of this delicate area involves removing failed gypsum render and replacing it with a new, more stable render that allows the walls to transmit moisture without damage. Careful scientific analysis and study were carried out on the original masonry including bedding mortars potentially dating from before the 1990s restoration, exterior and interior renders, and flooring material. 26 mortar and plaster samples, and one bedding sand sample, were submitted to a consultant laboratory for analysis. Mortar analysis was carried out according to BS4551: Part 2: 1998, whereby binder and insoluble aggregate proportions are determined through titration and acid digestion [1]. In addition, six of these samples were submitted for further petrographic analysis according to ASTM

C856 – 04, whereby thin sections are prepared in fluorescent epoxy resin and examined microscopically under fluorescent light [2].

The initial chemical analysis showed mainly gypsum : sand mixes (Table 1). However, chemical analysis alone is not able to distinguish between Portland cement and lime binders or between hydraulic and non-hydraulic limes. Petrographic analysis, however, revealed the presence of lime, which in some samples was in higher proportion than gypsum. There is very little insoluble residue remaining in most of the Delma samples following chemical analysis. This suggests that the aggregate is comprised primarily of ground soluble material such as carbonates and sulphates of calcium such as limestone, gypsum or shells, rather than silicate based sand. This is to be expected of sand sourced locally to the shore. Due to the use of soluble aggregates, the exact proportions of binder to aggregate were difficult to identify but the binder is in much higher proportion. The conclusion was that typical mortars contain gypsum, lime, and an aggregate of gypsum rock and/or limestone [3].

Because much damage to the buildings was caused by gypsum, which expands with moisture, lime was favored as a more stable conservation material. Both hydrated and natural hydraulic lime was sourced within the UAE. To determine the right mix for the new render, over 20 different samples were installed on site employing various mixes of lime (*noura*), sand and *sarooj,* a traditional waterproofing material produced in the UAE and Oman. *Sarooj*, which is a mixture of fired clay and organic materials, is added to the mortar as a pozzolan to improve strength and decrease setting time. Multiple render samples were installed which allow conservators to test factors such as strength, porosity and bond with the wall, and thus be able to select the best mix for conservation works. The selected render will be used for conserving the damaged lower walls, repairing cracks, and potentially re-rendering the building exteriors. In addition, new drainage systems, termite treatment, and repair of exterior facades and roofs will ensure the long-term stability of the buildings.

2.2. Qalat Umm Hisn

Figure 3: Qalat Umm Hisn in undated historic photograph (circa mid-20th century). Source: Center for Documentation and Research (CDR), Abu Dhabi.

Figure 4: The site in 2011 during laser scanning. Source: photograph by author.

Qalat Umm Hisn is a ruined stone tower at the western end of Liwa oasis, near the mahdar of Arada. The UK and Saudi Memorials of 1955, drawn up at the time of the Buraimi dispute, state that the tower was built by the Manasir tribe in the early years of the Qatar-Abu Dhabi war (1883 – 1890) [4]. Originally consisting of a round tower between two low stone enclosures, today, the structure is largely collapsed with only one section of the tower wall standing (Figs. 3 & 4). Structural assessment carried out in 2010 on the standing wall highlighted the urgent need for stabilization through crack stitching, wall capping and repointing of eroded mortar joints.

Addressing the building's conservation, ADACH conducted materials analysis to understand the type of mortar used in the building and to plan appropriate repairs. Analysis sought to answer basic questions regarding traditional materials used in a remote area with few natural resources. Would the mortar be based on lime or gypsum, and what is the material source and method of production? Three mortar samples were taken from different areas of the remaining tower wall and surrounding low walls. Analysis techniques included optical microscopy to examine resin-embedded cross sections, and scanning electron microscope/energy dispersive X-ray spectroscopy (SEM-EDX) and X-ray diffraction (XRD) to determine elemental composition as well as mineral phases and crystalline compounds (Figs. 5 & 6). SEM investigation was performed on a Quanta 200 (FEI Company), equipped with a Wolfram-cathode with a Quantax microanalytical EDX add-on (Bruker AXS). X-ray diffractograms were recorded by a Philips PW 1729 equipped with a copper tube (30KV, 30 mA) on silicon single-crystal sample carriers [5].

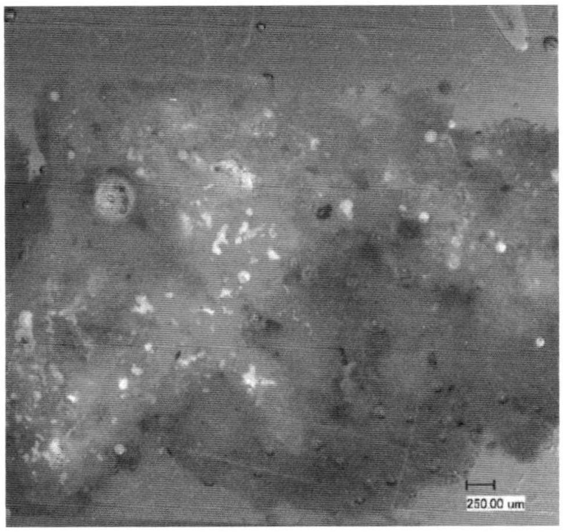

Figure 5: Resin-embedded mortar cross section under microscope. Source: Stefan Simon.

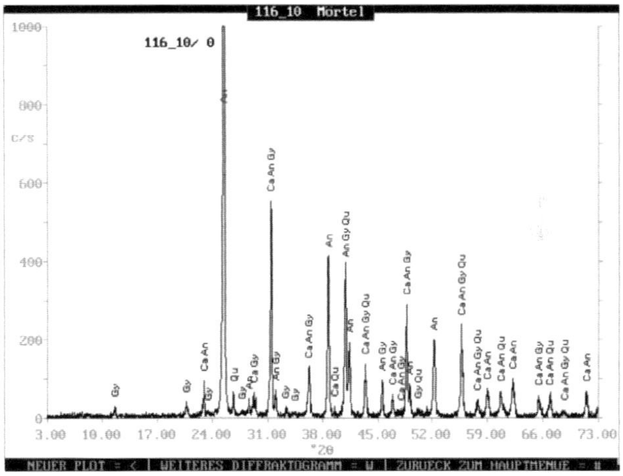

Figure 6: XRD results showing predominance of Anhydrite. Source: Stefan Simon.

Analysis revealed a composition of mainly Anhydrite ($CaSO_4$), which is the thermodynamically stable phase of gypsum ($CaSO_4{\cdot}2H_2O$) when it is heated above 42 degrees Celsius. The mortar is homogenous with small percentages of fine quartz, feldspar and calcite aggregate [5]. Anecdotal evidence provided by restoration craftsmen working in Liwa further clarified the source and method of mortar production. They state that the mortar is based on local gypsum stone which is mined from the valley floors between the dunes, burned using palm-wood fuel, then crushed to a powder and mixed with sand and brackish well-water to form a mortar [6]. The analysis provided information critical for developing compatible repair mortars such as the composition of the binder and the volume, size and type of aggregate. It also confirmed the use of gypsum stone suggested by the oral tradition, and will help conservators to select appropriate local materials for future testing of mortars and renders during conservation treatment.

2.3. Al Hayla Tower

Al Hayla tower is an unrestored stone building located on the western side of Liwa Oasis near the village of Marya al Gharbiya (Figs. 7 & 8). The tower is thought to have been built in the 19th century during the wars between Liwa and Ajman [7]. Constructed of local stone and traditional gypsum-based mortar, the tower may have served defensive purposes, been a mark of land ownership, or protected a local water source. As the only intact stone building in Liwa Oasis, Hayla Tower is an important record of stone building techniques in Liwa and a significant part of Abu Dhabi's heritage.

Figure 7: Hayla Tower in undated historic photograph (circa mid-20th century). Source: CDR.

Figure 8: Hayla Tower in 2011 during installation of retaining straps. Source: photograph by author.

When first surveyed by ADACH, the tower was in extremely dangerous condition. With its roof and floors gone, the walls were left vulnerable and large cracks in the masonry had developed that threatened to cause total collapse of the structure. In 2009, support scaffolding was installed and the largest cracks were stitched with stone and mortar (Fig. 8). This was followed by additional structural stabilization including the installation of retaining straps around the tower exterior. A conservation management plan was initiated which guides conservation decisions and included documentation through 3D laser scanning and laboratory analysis of mortars.

Mortars analysis was conducted using the same techniques as at Delma Island - a combination of wet chemical analysis and petrography. Analysis of 6 mortar samples showed that the binder is gypsum, while insoluble residue content is high which confirms that the aggregate is predominantly silicate based, which is to be expected given the tower's location in the sandy desert far from the coast (Table 2). Petrographic examination identified some limestone particles in the samples as aggregate, but from the chemical analysis it can be inferred that the percentage of limestone is low, ranging from 5.5% to 13.5%. The ratio of sand : gypsum is consistent across the samples at around 1:0.5 [8].

CHEMICAL ANALYSIS	H1	H2	H3	H4
Insoluble Residue	33.65	34.82	34.13	33.07
Soluble Silica, SiO_2	2.63	1.10	0.92	0.96
Acid soluble Alumina, Al_2O_3	0.17	0.13	0.07	0.09
Acid soluble Iron, Fe_2O_3	0.05	0.02	0.02	0.02
Acid soluble Calcium, CaO	24.41	23.92	25.74	25.62
Acid soluble Magnesium, MgO	1.86	0.51	1.11	0.81
Acid soluble Sulphate, SO_3	29.27	28.87	26.45	32.02
Loss on Ignition	7.92	10.11	10.89	7.28
Total	99.96	99.48	99.33	99.87

Table 2: Chemical analysis results of four mortar samples showing gypsum ($CaSO_4 \cdot 2H_2O$) and insoluble residue (sand). Source: DIA.

While gypsum was identified as the main component of the building's mortar, conservators have chosen lime as a repair material over gypsum for its stability against moisture and salts. Trials of lime-based repointing mortars, injection grout, and a shelter coating were carried out in June 2011. As at Delma, various mixes of lime, coarse sand and the pozzolanic *sarooj* were tested on a small area of the tower. After evaluation of the trial mortars, taking into consideration factors such as strength, adhesion, color and porosity relative to adjacent materials, full conservation treatment will be implemented followed by fencing, signage, and interpretive features to make the site accessible to visitors.

IOP Conf. Series: Materials Science and Engineering **37** (2012) 012004 doi:10.1088/1757-899X/37/1/012004

3. Earthen building materials

3.1. Hamad bin Hadi al Darmaki House, Al Ain

The Hamad bin Hadi al Darmaki House is an earthen structure located in Hili Oasis in northern Al Ain. Comprising several rooms and a tower situated around a courtyard, the site is enclosed with high earthen walls that rise several meters above the surrounding oasis (Fig. 9). Materials analysis including X-ray Diffraction (XRD), X Ray Fluorescence (XRF), and soil mechanics was carried out in 2007 to characterize the mud bricks and understand different phases of construction. Material sampling was carried out in two stages: first, 9 samples were taken for various tests, including 5 mud-brick samples, 2 plaster samples, 1 mortar sample, and 1 stone sample; the second stage was a continual study of the mud-bricks during which two more samples were taken. Since a number of different sizes of mud bricks were found, this study helped to identify the right size of mud bricks to be used in conservation.

Figure 9: Hamad Bin Hadi al Darmaki House, 2011. Source: Salman Muhammad Ali.

A sample of local clay-bearing soil from the Al Saad area was tested as well, so that it could be compared to the mud-brick samples for use in restoration. Based on the poor condition and cracking seen in the historic material, it is evident that the soil originally used to make the mud-bricks of Bin Hadi House is not of good quality. As indicated in the XRD analysis (Table 3, left), the mud-brick samples BH-01, BH-05, and BH-12 contain large percentages of the clay mineral montmorillonite. Montmorillonite is highly absorptive and increases greatly in volume with water. The large amount of montmorillionite in the mud-bricks would result in expansion and contraction of the clay surface causing cracks to form. In addition, montmorillonite loses viscosity over time, which results in the clay surface breaking down and powdering.

Sample No.	Major phase	Minor phase
BH-01	Calcite (29%), dolomite (25%), montmorillonite (20%)	Quartz (10), orthoclase (8%), albite (7%)
BH-03	Calcite (29%), dolomite (25%), montmorillonite (20%)	Quartz (10), orthoclase (8%), albite (7%)
BH-04	Gypsum (90%)	Quartz (8%)
BH-05	Calcite (31%), montmorillonite (23%), dolomite (14%), quartz (12%)	Chlorite (7%), orthoclase (6%), albite (5%)
BH-12	Calcite (30%), quartz (18%), dolomite (15%), montmorillonite (14%)	Albite (8%), chlorite (7%), orthoclase (5%)
Al Saad	Quartz (24%), montmorillonite (22%), calcite (14%), dolomite (11%)	Albite (10%), chlorite (10%), orthoclase (6%)

		BH-01	BH-03	BH-04	BH-05	BH-12	Alsaad
SiO_2	%	29.05	18.11	8.71	23.93	23.34	36.20
Al_2O_3	%	2.95	2.94	1.14	2.43	2.71	4.03
Fe_2O_3	%	3.34	3.42	1.58	2.83	4.02	5.99
CaO	%	29.17	29.24	25.04	32.80	27.01	17.94
Na_2O	%	0.63	0.28	0.09	0.36	0.43	0.30
K_2O	%	0.95	0.84	0.26	0.78	0.85	0.95
MgO	%	8.05	11.29	4.52	8.94	8.66	13.64
TiO_2	%	0.215	0.181	0.081	0.163	0.184	0.285
MnO	%	0.056	0.063	0.023	0.050	0.050	0.095
P_2O_5	%	0.043	0.060	0.075	0.056	0.049	0.065
LOI	%	24.62	32.32	20.09	26.88	22.86	19.66
SO_3	%	0.145	0.135	37.900	0.160	0.065	0.173

Table 3: Results of XRD (left) and XRF analysis (right). Source: Ali Malekabbasi / Zarazma Testing.

Based on the analysis results and the examination of area soils it is evident that sand-lime structures dominate in the Al Ain area, with only trace amounts of clay minerals. This has resulted in a quality varying from average to poor for the soil used in mud-bricks. The most significant information found through XRF analyses (Table 3, right) is the very low amount of aluminum oxide (Al_2O_3), which is limited to only 2 to 3 percent in the samples. The acceptable amount of aluminum oxide in mud-bricks that perform well in tests is around 8 percent. The size grading test, used to separate and quantify the contituents of soil, also confirmed these results. Thus no better condition could be expected of the historic mud-bricks of Bin Hadi House considering the relative percentage of these elements [9].

The materials analysis at Bin Hadi house also confirmed the suitability of the clay from the Al Saad area for conservation, as it contained comparable quantities of montmorillionite and other key minerals that indicate compatibility with the historic bricks. The clay was taken from the source and combined with straw and sand to make mud bricks and plasters for repairing damaged areas. Many of these repairs have been in place for over five years with little damage or alteration, demonstrating their compatibility and the success of the analysis in determining appropriate conservation materials.

3.2. Hili 17

Figure 10: House 1 in the early 1990s. Source: DAA n. 2

Figure 11: Detail of deterioration. Source: author.

Hili 17 is an Iron Age II period archaeological site located in Hili Archaeological Park, Al Ain. The site comprises three earthen houses and numerous hearths built over several centuries of occupation. It has been carbon dated to ca. 1000 BCE and is significant as a possible pottery production center [10, 11]. Excavated from 1991-94, Hili 17 has been exposed for over 20 years and has deteriorated significantly with many areas of collapse and general loss of definition from erosion (Figs. 10 & 11). The rapid rate of deterioration is thought to be related to cracks formed by swelling and shrinkage of the white clays used for mud bricks and plaster, and to salts in the walls and surrounding sand. ADACH's conservation approach for Hili 17 has been to document the site through laser scanning, analyze the clay materials to understand deterioration mechanisms, and in the short-term begin emergency conservation treatments while studying the implementation of a long-term protective shelter. Initial materials analysis included microscopy, XRF, XRD, and SEM.

Microscopic analysis revealed a number of factors important to understanding the site's decay process; microscopy showed an extremely fine-grained material with few large particles. The sand fraction is limited to a maximum grain size of ca. 0.3mm, and the majority of the material is silt or clay sized fractions. Such fine grained material is much more susceptible to cracking than earthen material made with graded aggregates. Microscopy also showed elongated voids, indicating that the mud-bricks and plaster may have originally been reinforced with vegetable fiber (a practice that continues today) but that the fibers have disintegrated over time (Fig. 12).

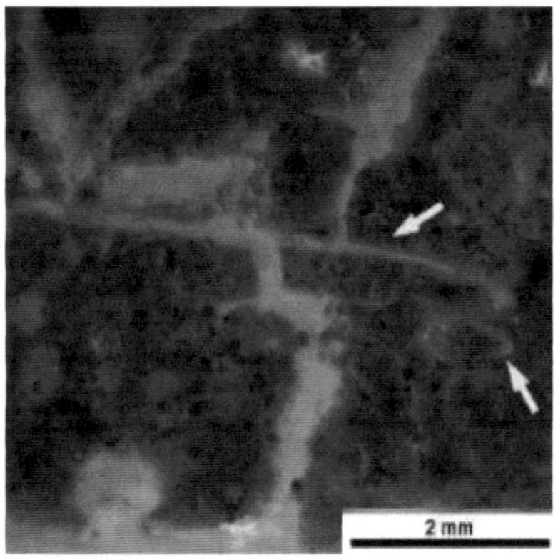

Figure 12: Microscopic image showing elongated voids under UV light. Source: Christof Ziegert, ZRS.

XRD results (Fig. 13) included minerals such as dolomite, kaolinite, calcite and albite. It also revealed the presence of swelling clays such as palygorskite, which is often associated with smectite [12]. Smectites are fibrous clays with small particle sizes and large internal surface areas. They have negative layer charges which allow them to absorb several times their weight in water. The presence of this type of clay, combined with precipitation and atmospheric moisture causes expansion and contraction of the clay surface and the subsequent cracks and deterioration seen at Hili 17. XRD also revealed a high concentration of halite (NaCl), or sodium chloride. Further testing using electrical conductivity showed nearly double the amount of soluble salts at the bases of the walls as at the tops. The high salt content may explain the deterioration of the lower walls, as wetting and drying causes cyclical salt dissolution and re-crystallization that erodes the wall.

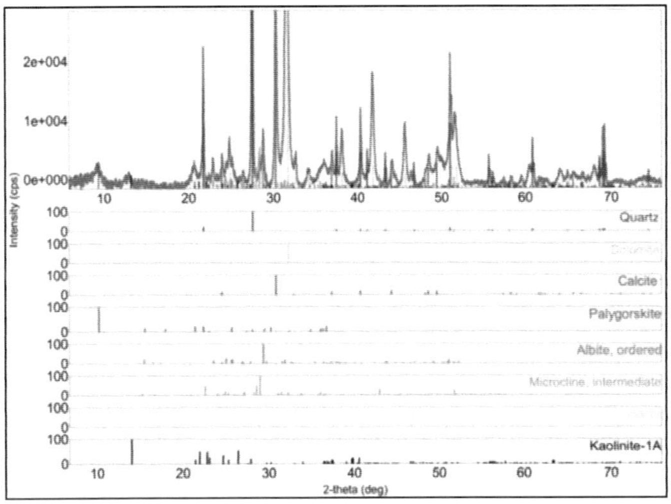

Figure 13: Results of XRD analysis showing presence of swelling clay (palygorskite) and salts (halite). Source: Cristof Ziegert, ZRS.

After analysis of the historic clay, XRF was used to identify a suitable material for conservation interventions. Clay-bearing soil from six different areas within the park was sampled and compared with the original material for color and soil mechanics including rate of water absorption and percentage of sand, silt, and gravel. Samples from the most promising natural clay layer, located about 20 meters east of the site, were submitted for XRF analysis at the University of Sharjah and compared with samples of mud bricks from Hili 17. Four samples were powdered and analyzed using a Horiba XGT-7200 micro-XRF microscope.

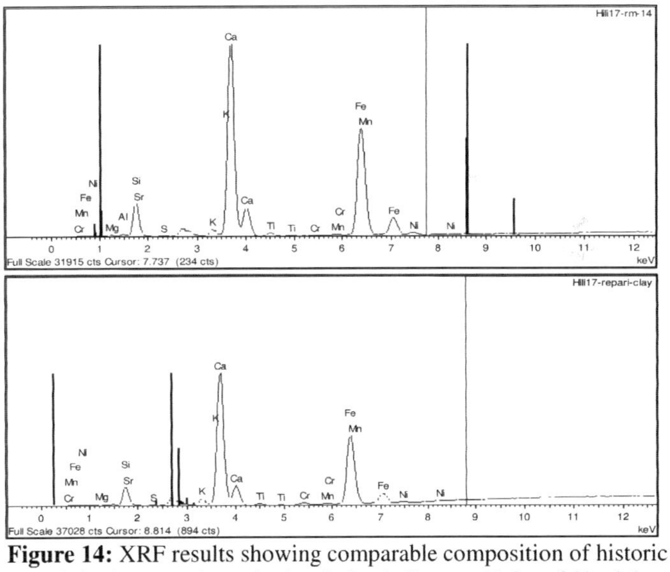

Figure 14: XRF results showing comparable composition of historic material (above) and repair clay (below). Source: Univ. of Sharjah.

XRF Analysis (Fig. 14) showed comparable amounts of key elements calcium, iron, magnesium, and silicon, and indicated that the new clay was indeed similar and compatible with the original. Based on the analysis results, the identified clay was mined from the source layer and brought to the ADACH workshop for further field testing. This white clay is used to make mud bricks, plasters and grouts for the repair of damaged areas.

4. Conclusion

The analyses presented in this paper have revealed important information regarding both the composition of traditional building materials in Abu Dhabi as well as their limitations. The studies of mortars have shown that in remote areas such as the Liwa Oasis, stone buildings were constructed using mortars based on readily available stone with desert sand aggregate. This evidence is corroborated by oral testimony, which points to the burning of local gypsum rock for the creation of mortars. At Delma Island, analysis results show that lime was used in mortars and that carbonate aggregates are more common on the coast than the sand aggregate of the desert interior. However, the results of the Delma island mortars are complicated by later restoration and further analysis of historic coastal buildings is needed to understand the sources and use of lime-based mortars on the coast.

The studies have also highlighted the limitations of local historic building materials, particularly in regards to earthen buildings, which are constructed with sandy clays and problematic minerals such as montmorillonite and smectite. In addition, analysis revealed concentrated soluble salts, a problem that affects most masonry materials in Abu Dhabi and causes rapid and significant damage. The combination of salts and poor materials leads to a high rate of deterioration that must be addressed through conservation with stable and compatible materials. While gypsum is common in the regions historic mortars, its poor performance in the presence of moisture and salts has led ADACH to use more stable materials in conservation such as lime and *sarooj*. Similarly, the weak and sandy clays used historically in earthen buildings are not suitable for conservation, and more suitable clays have been chosen for repairs. Through analysis, ADACH has been able to gain a more thorough understanding of Abu Dhabi's built heritage, diagnose decay mechanisms, and make informed decisions regarding conservation and management.

Acknowledgments
Ali Malekabbasi and Ibrahim Khalil El Lababidi, ADACH; Dr. Cristof Ziegert, Ziegert Roswag Seiler (ZRS) Architekten Ingenieure, Berlin; Dr. Stefan Simon, KDC Konservierung & Denkmalpflege Consulting, Olching, Germany; Simon Malam and Chris Wilson, Donald Insall Associates, London; Dr. Najeh Jisrawi, University of Sharjah, UAE.

References

[1] British Standards Institution (1998). *Methods of Testing Mortars, Screeds and Plasters - Part 2: Chemical Analysis and Aggregate Grading.* BS4551- 2:1998. 389 Chiswick Rd. London.
[2] ASTM International (2010). *Standard Practice for Petrographic Examination of Hardened Concrete.* ASTM C856 – 04. West Conshoocken, PA: ASTM.
[3] Donald Insall Associates (2010). *Delma: report on sample testing carried out for Abu Dhabi Authority for Culture and Heritage.* Unpublished report prepared for ADACH.
[4] Sheehan P (2010). *Report on site visit to Qalat Umm Hisn.* Unpublished internal report.
[5] Simon S (2010). *Mortar Analysis: Umm Hisn.* Unpublished report prepared for ADACH.
[6] Interview with Mr. Abdul Salam, Senoir ADACH craftsman in Liwa. Conducted by Ali Malekabbasi and Benjamin Marcus, June 2011.
[7] Donald Insall Associates (2010). *Al Hayla Tower: Conservation Managment Plan.* Unpublished report prepared for ADACH.
[8] Donald Insall Associates (2010). *Liwa: report on sample testing carried out for Abu Dhabi Authority for Culture and Heritage.* Unpublished report prepared for ADACH.
[9] Malekabbasi A and Sheehan P (2011). *Hamad bin Hadi al Darmaki House and the cultural landscape of Hili Oasis, Al Ain, UAE* (Forthcoming).
[10] Magee P (2003). *New chronometric data defining the Iron Age II period in south-eastern Arabia.* Proceedings of the Seminar for Arabian Studies 33: 1-10. Archaeopress: Oxford.
[11] Yasin al Tikriti W and Hadou M.M. (2001b). Hili 17, in Yasin al-Tikriti, W. and Mouton, M. (eds.), *The Architectural Remains of the Iron Age Sites in the United Arab Emirates and Oman.* Documents d'Archeologie de l'Arabie (DAA) n. 2, CD-ROM multimedia. Maison de l'Orient Mediterraneen, Lyon.
[12] Ziegert C (2010). *Materials analysis: Hili 17.* Unpublished report prepared for ADACH.

IOP Conf. Series: Materials Science and Engineering **37** (2012) 012005 doi:10.1088/1757-899X/37/1/012005

Role of XRF in the restoration of a prominent architectural monument at the site of Mleiha

Eisa yousif[a] and Atta Attaelmanan[b]

[a] Directorate of Antiquities, Sharjah, UAE

[b] University of Sharjah, Sharjah, UAE

Abstract
The fort at Mleiha is one of the most well known archaeological sites of its period (3rd c. BC to 3rd c. AD) in the United Arab Emirates and has provided much valuable information. The period of its construction and use coincided with the Parthian period, during which there were extensive commercial contacts between the UAE and Mediterranean cities, Egypt, Mesopotamia, Southwest Iran, Pakistan, East Africa, and the southern Arabian Peninsula. It was restored using local materials, and traditional construction methods.
The objective of this study was to utilize XRF techniques for the identification of the elemental composition of building material used to construct an archaeological fort, and to compare it to composition of the restoration material.
Results of the study show strong correlation between the original fort building material and those used for restoration.

1. INTRODUCTION

The fort at Mleiha is one of the most well known archaeological sites of its period (3rd c. BC to 3rd c. AD) in the United Arab Emirates and has provided much valuable information. The period of its construction and use coincided with the Parthian period, during which there were extensive commercial contacts between the UAE and Mediterranean cities, Egypt, Mesopotamia, Southwest Iran, Pakistan, East Africa, and the southern Arabian Peninsula.
Although the fort was constructed shortly after the end of the Iron Age, around 300 BC, the inhabitants of Mleiha appear to have no relationship with the Iron Age inhabitants of the area. It is considered possible, therefore, that they may have originated from Arab Bedouin groups that formerly resided in the centre and northern parts of the peninsula. (Hellyer 1999, Potts 2001).
Excavations in Mleiha have uncovered a number of houses, cemeteries and animal burials, as well as other large buildings that may also have been fortified, besides the Mleiha fort itself (Jasim 1999).

1.1 Summary of Excavations

The Mleiha fort was first identified in 1990, as a result of the digging of a water pipeline adjacent to the highway that formerly ran directly across the site. Remains of walls and a fragmentary coin mould were among the key finds. Test trenches were initially excavated by the French Archaeological Mission (Mouton et al 1997, Mery and Mouton 2011), followed by detailed excavations in 1992.

These French excavations indicated that the fort had been built on a low, east-facing hill, and that it was constructed entirely of mud-brick. It has been dated to the First and Second Centuries BC.

IOP Conf. Series: Materials Science and Engineering **37** (2012) 012005 doi:10.1088/1757-899X/37/1/012005

The fort is square in shape, with eight towers, one at each corner and the others in the centre of each wall. The walls, of considerable thickness, are approximately 55 meters in length, with the entrance being located in the middle tower on the eastern wall. The interior of the fort takes the form of a central courtyard, with a number of small rooms attached to the walls, the walls of the rooms being 80 centimeters in width, and with doorways being 90 and 120 centimeters in width (Figure 1).
The excavations were undertaken in two phases. The first phase examined that part of the fort which was not covered by the tarmac road, while the second stage of excavation, of, the central part of the fort, took place following the diversion of the tarmac road, in accordance with directives from His Highness Dr. Sheikh Sultan bin Mohammed Al Qasimi, Ruler of Sharjah, to preserve the site and to facilitate excavation .

Figure 1. A photograph of the Mleiha fort showing the fort's foundations and its position relative to the main road. The road was moved on the orders of HH Dr Sultan Al Qassimi, ruler of sharjah, so as to preserve the archaeological site. (*Copyright: Diractorate of Antiquities, Sharjah*)

2. Mleiha Fort Restoration

Following the excavation of the fort carried out by the French Archaeological Mission, it was decided to restore and preserve the surviving remains, in particular those parts damaged through erosion or by their proximity to the former tarmac road. This work was carried out, in several stages, by the Sharjah Archaeological Team, making use of the plans prepared during excavation and of other data relating to the building.

2.1. Stage I

This included excavation and cleaning work next to the walls, to look for any other buildings that might be attached to its outer walls. The following findings were made:
2.1.1 A single room attached to the fort and situated between the North-East tower and the East Tower, with a single entrance. The wall of the room, which is linked to the two towers, and is 12.57 centimeters in width, with an entrance of 95 centimeters. This appears to have been added to the fort after its original construction. The room is divided into two equal parts, with a dividing wall of 2.90 meters in length, with an entrance of 99 centimeters width. A small hearth is present in one of the rooms, which may have been used both for cooking and heating .
2.1.2 A water channel has been excavated in the wall joining the South and South-West Towers, this being lined with stones at its base and on its sides to prevent erosion by the flow of water. It leads into one of the interior rooms, in which the French Archaeological Mission had previously identified another channel through the interior wall to the main internal courtyard of the fort .

2.1.3 A number of stones were found close to the South-East Tower, one of which has a carved image of a standing person. The stone is of sedimentary rock typical of the Jebel Faya and Jebel Mleiha formations.

2.2. Stage II

During this stage, of the restoration work, research was undertaken to determine the primary sources for the mud-bricks and pebbles used in the construction of the fort.
Samples of mud were collected from a number of wadis and farms close to the site, but these appeared to be unsuitable or the manufacture of mud-bricks, since they rapidly cracked once dried.
Eventually, an area containing old muddy sediments was identified east of Jebel Mleiha, in the Khatim region . This appeared to represent a now-dried up lake-bed from a period of higher rainfall
The thickness of the muddy sediment in this lake bed is between 15 and 20 centimetres. Sample mud-bricks prepared from the sediment showed a resistance to cracking, while they also had the same properties as the mud-brick from which the fort was constructed.
Sources of small pebbles were located approximately 700 metres from the southern side of the fort and 500 metres to the north-east of the fort.
After testing the lake bed sediment, production of mud-bricks for the reconstruction was commenced.

2.3. Stage III :

This stage included the preparation of the mud-brick and the reinforcing of the surviving foundations of the fort's walls. The soil and mud was not subjected to any laboratory test.
It appears that the mud-brick used to construct the original fort was made by mixing the mud with small pebbles and then drying them on mats of palm fronds. Impressions of palm fronds have been identified on some of the bricks. The bricks used were of two sizes, one of 40 centimetres length, 40 centimetres width and 10 centimetres thick, and the other of 40 cm by 20 cm by 10 cm. Bricks of these sizes have also been noted in other buildings in the Mleiha area
On the basis of these specifications, the mud-bricks were made as follows:

2.3.1. Mixture Preparation process :

Gravels / small pebbles were collected from an area near the fort and were then sieved to remove fragments of glass. Mud was then added, at a 3:1 ratio, i.e. one load of gravel being mixed with three of mud.
A basin was then made, which was then filled with water. The gravel/mud mix was then added, with the resulting mix being left to stand for a whole day.

2.3.2. Preparing molds, adding mixture and making mud-bricks:

Wooden moulds were made of the same sizes as of the ancient mud-brick .
The amount of time during which the mixture of gravel, mud and water was left in the basins varied from season to season, being left for 4 to 5 days in winter, for the larger bricks, and 2 days in the summer, because of varying rates of evaporation .
The wooden moulds were then removed from the mud-bricks, which were left to dry further, for 3 days in the winter and 2 days in the summer .
The mud-bricks were then stood on end for further drying, for 2 days in winter and a day in summer .
The above process indicates that temperature, and the speed of the consequent process of watert evaporation, is of importance for the production of the mud-bricks.

2.3.3. Re-constructing Foundations and Restoration

The foundation level was restored prior to the laying of the mud-brick.
This was required because some parts of the foundations had been damaged by bulldozers during the process of removing the old tarmac road that formerly ran across the centre of the site.
Careful examination showed that the original foundations had been laid on the natural surface. These were restored .

2.4. Stage IV

After erecting the walls, the interior floors of the fort, both of the internal rooms and of the courtyard, were covered with a thin layer of mud, on which machinery could work during the process of erecting the cover over the whole structure, thus avoiding damage to the original floors. This mud, obtained from wadi silt and then cleaned of extraneous material, was mixed with sand on a 2:1 ratio. This temporary floor was of around 5 centimetres in thickness .

3. XRF Analysis of building material

Samples from the fort's building material were analyzed using an X-ray analytical microscope (XGT 7200, Horiba, Japan). XRF analysis were performed using an intense X-ray beam 1.2 mm in diameter, generated from an Rhodium (Rh) X-ray tube operating at 50 kV and 0.5 mA. Four batches of samples were collected. Two batches from the original fort material (O), and the other two batches from the restoration material (R). Each batch contained three samples in powder form. Samples from the original material was collected from drills made in the core of the fort bricks, while restoration material samples were collected from the lake bed sediment used to manufactur the restoration bricks.
The system was calibrated using a brick clay standard (SRM 679) which was analyzed simultaneousely with the four samples. Measured concentrations were very close to the reference values (fig. 2) such that the fitting linear curve had a slope of 0.996 and a fitting factor R^2 of 0.9558.
Each sample was analyzed under vacuum for 1000 seconds so as to improve the probability of detecting the trace elements. Collected spectra were deconvoluted using quantification software, that caloculates elemental concentrations according to the fundamental parameter method.
Element oxide concentrations for the 12 samples shows the presence of the same elements in almost equal quantities. Silicon (Si), Calcium (Ca) and Iron (Fe) constitute the major elements, while Magnesium (Mg), Aluminum (Al), Sulfur (S) and Potassium (K) are the minor elements. Even the same trace elements are present in all the samples in almost equal amounts.

4. Conclusions

Correlation of the composition of the original fort material and the restoration material shows that the two materials are closely matched. A plot of the elemental concentrations for the original material versus the restoration material (fig. 3) has a slope of 1.01, which indicates a very good correlation between the two materials.
Variations in the concentration levels between the two materials could be explained by the simple fact that changes are expected during the manufacturing of the mud-bricks. However, despite the variations the concentrations, our results indicate that the source of the original building material used in the construction of the fort, almost two centuries ago, was the identified area east of Jebel Mleiha, in the Khatim region.

5. Acknowledgment

Our thanks to Dr Sabbah Jasim, for the permission to use the fort's photograph (fig. 1).
The X-ray fluorescence analysis was conducted at the National X-ray Fluoresnces laboratory, Sharajah, UAE.

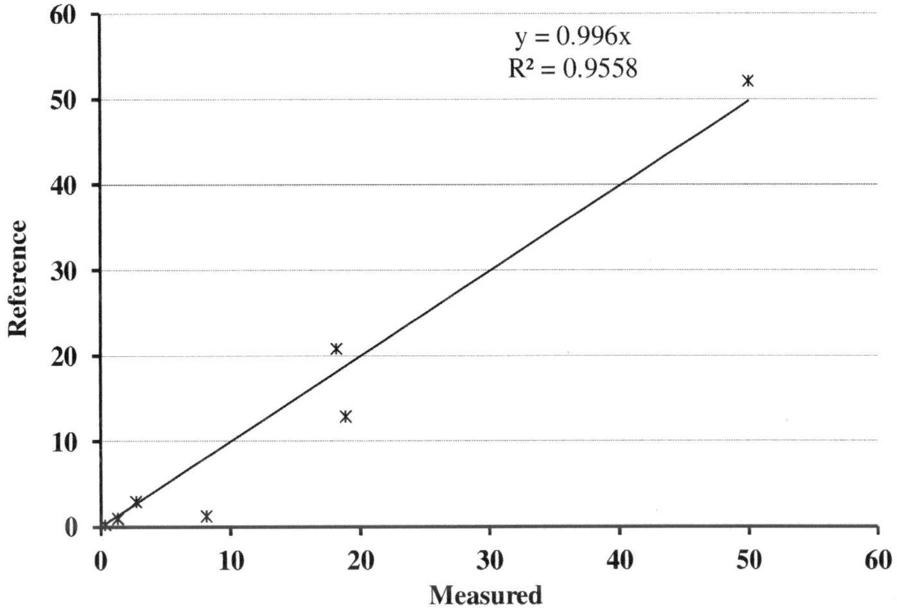

Figure 2. Calibration curve showing reference element concentrations versus measured values.

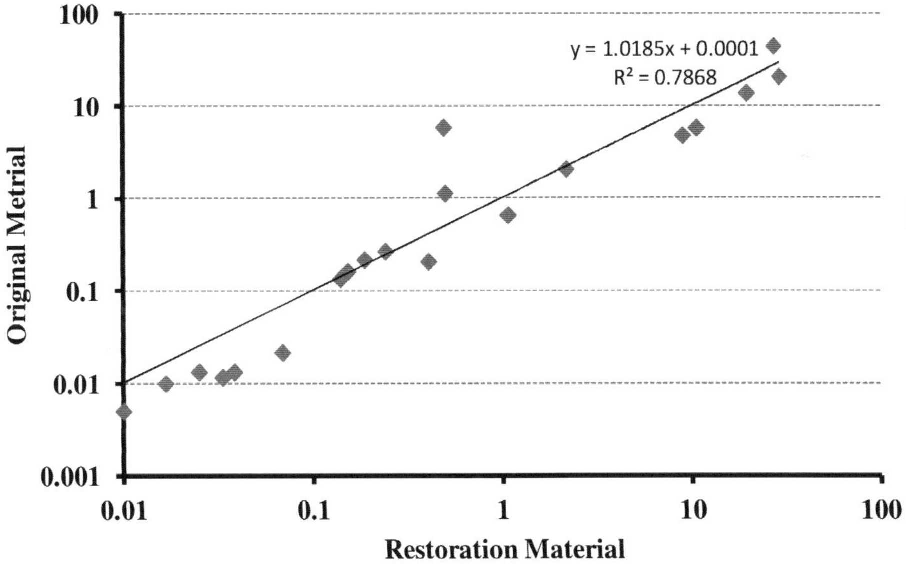

Figure 3. showing average element concentrations for the original fort material versus the restoration material

IOP Conf. Series: Materials Science and Engineering **37** (2012) 012005 doi:10.1088/1757-899X/37/1/012005

Table 1. Average element concentration values (mass %) for the 12 samples from the restoration (R) and original (O) building materials used in the Mleiha fort. Error values are 3σ. Elements with concentrations less than 0.1% (Cu, As, Rb and Y) are not shown here.

	R1	R2	O1	O2
MgO	11.64±0.74	9.31±0.78	5.22±0.62	6.30±0.59
Al_2O_3	10.66±0.21	7.11±0.15	4.67±0.14	4.88±0.17
SiO_2	32.13±0.21	25.87±0.12	20.28±0.1	21.10±0.15
SO_3	0.45±0.03	0.53±0.03	5.71±0.03	5.92±0.03
K_2O	2.13±0.03	2.22±0.03	2.09±0.02	2.03±0.02
CaO	23.45±0.23	30.52±0.12	45.22±0.13	43.29±0.11
TiO_2	0.98±0.02	1.15±0.02	0.72±0.02	0.60±0.02
Cr_2O_3	0.12±0.01	0.25±0.01	0.16±0.01	0.27±0.01
MnO	0.36±0.01	0.45±0.01	0.20±0.01	0.22±0.01
Fe_2O_3	17.02±0.15	21.22±0.08	13.97±0.05	13.60±0.04
NiO	0.20±0.01	0.28±0.01	0.28±0.01	0.26±0.01
SrO	0.42±0.01	0.57±0.01	1.120.01	1.15±0.01
ZrO_2	0.15±0.001	0.15±0.001	0.13±0.001	0.19±0.001
SnO_2	0.13±0.016	0.14±0.026	0.14±0.02	0.13±0.02

References

1. Hellyer P., 1999. Hidden Riches - An Archaeological Introduction to the United Arab Emirates. Union National Bank, Abu Dhabi. 99 (Arabic version) .

2. Jasim S. A., 1999. The excavation of a camel cemetery at Mleiha, Sharjah, UAE. Arab. Arch. Epig. 10: 69 – 101.

3. Mery S., Mouton M., 2011. French archaeological expedition in the United Arab Emirates: Excavations at Mleiha, Sharjah Antiquities Annual Magazine. Vol 13, 64 – 83.

4. Mouton M., Mokaddem K., Garczynski P., 1997. Excavations at Mleiha: 1990 and 1992 campaigns. Sharjah Arachaeology 6th report. 24 – 43.

5. Potts D. T., 2001. Before the Emirates: an Archaeological and Historical Account of Developments in the Region c. 5000 BC to 676 AD. Book chapter in United Arab Emirates: A new perspective. Edited by Al Abed I., Hellyer P., Trident Press Limited, London, UK. ISBN 1-900724-47-2, pp 28 – 69.

Integration of μ-XRF, and μ-Raman techniques to study ancient Islamic manuscripts

N. M. Hamdan[1,2], H. Alawadhi[2,3], N. Jisrawi[2,3]

[1]Physics Department, The American University of Sharjah, Sharjah, United Arab Emirates
[2]National X-ray Fluorescence Laboratory, Sharjah, United Arab Emirates
[3]Department of Applied Physics, University of Sharjah, Sharjah, United Arab Emirates

E-mail: nhamdan@aus.edu

Abstract: Systematic characterizations of inks used in three ancient Arabic manuscripts were performed using energy dispersive X-ray fluorescence (EDXRF) and RAMAN spectroscopy. Spatially resolved element distributions obtained by area scans in addition to spot analysis provided detailed information about the manuscripts. RAMAN spectroscopy, on the other hand provided decisive information about the compounds used in various color pigments. Different compounds were found in the red ink in each manuscript: HgS (cinnabar) in one manuscript, red lead Pb_30_4 (minium) in the second, and a mixture of the two in the third manuscript. The green pigment was Cu based and it is probably malachite ($Cu_2CO_3(OH)_2$), while the black ink pigment in the three manuscripts is iron gall (Fe_2SO_4 plus gallottanic acid). Analytical results and the context of the manuscripts were used to extract information about the authenticity and origin of the manuscripts. For example, the use of cinnabar as a red ink pigment found in the third manuscript reveals that this manuscript attributed to Al-Gazali (11th century) is probably a copy from a later period, because cinnabar was first introduced in the 16th century.

1. Introduction

Identification of pigments in old Islamic manuscripts is important for conservation and restoration purposes, so is the detailed study of compounds used in the inks. In a preliminary investigation of a set of fragments from old Islamic manuscripts two complementary techniques were utilized. X-ray fluorescence (XRF) and micro x-ray fluorescence were used for elemental analysis. However, since the XRF technique can only identify elements heavier than Na, and can't identify the chemical structure, we used μ-RAMAN, a compound-specific technique, in an attempt to determine the chemical composition of various color pigments. The integration of the two techniques provides comprehensive information about the chemical composition of various pigments and the structure and manufacturing processes of ancient paper. In addition, such investigation may also provide information about the degradation of metallic pigments in ink used in ancient manuscripts [1-5].

We present preliminary measurements on three fragments from ancient Islamic manuscripts obtained from Juma Al Majid Center for Culture and Heritage in Dubai. The first fragment is a page from an unknown old Arabic poem (poet and date of writing also unknown); while the second is a page from a manuscript by Sirajudiin Omar Bin Ali known as "Qari' Al-Hidaya", a religious scholar who died in the year 826 A. H. in Baghdad. The manuscript contains answers to questions on various religious matters by the scholar. The manuscript is known as "FATAW's of Qari' Al-Hidaya". It is believed that the manuscript was written by Al-Hammam Bin Abdel-Wahid in 861 A. H. It is thought that the same manuscript was rewritten in 990 A. H. a version available at King Saud University Library in Riyadh, Saudi Arabia. The third fragment used in this investigation is the introductory page of a book entitled "Rules of Doctrines" by Imam Al-Ghazali (Abu Hamed Mohammad Ibn Mohammad Al-Ghazali (1058–1111 A.H.) as appears from the text on the fragment [6].

2. Experimental

The XGT 7200 Horiba μ-XRF system at the National X-ray Fluorescence Laboratory, Shrajah, was used for elemental mapping and for spot analysis of various color pigments of the three manuscripts. The instrument is equipped with 10 μm and 1.2 mm micro-capillary beams both of which were used in producing the mappings. The 1.2 mm beam was used for spot analysis at 50 keV x-ray energy. The Raman measurements were done using an AvaRaman system from Avantes. The excitation 785 nm laser with a power of 40 mW was focused down to a spot size of about 100 μm for the final measurement.

3. Results and discussions:

3.1 Manuscript 1

Figure 1 shows both sides of the fragment identified here as manuscript 1 annotated with the analyzed spots.

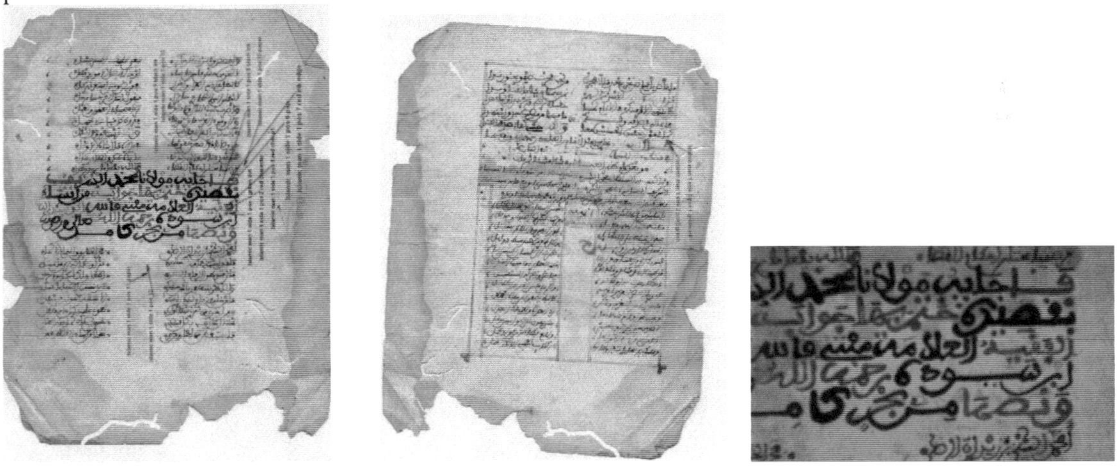

(a) **(b)** **(c)**

Figure 1: Manuscript 1: (a) side 1; (b) side 2; (c) mapping area of side 1

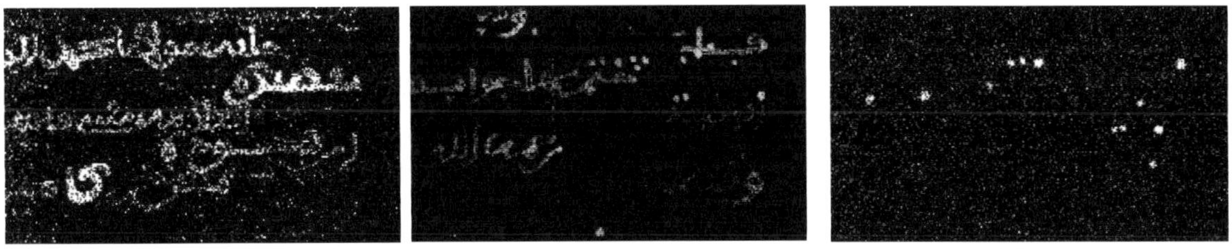

(a) **(b)** **(c)**

Figure 2: μ-XRF Elemental maps of manuscript 1 side 1: for (a) Fe; (b) Pb and (c) Cu

Figure 2 shows the elemental map for the three colors used in the manuscript: (a) the iron map corresponding to black ink, (b) Lead map corresponding to red ink and (c) Cu map corresponding to the green color ink. The maps were produced at a 10 μm beam with 50 keV x-ray energy. These results were also confirmed using spot analysis for various points shown in Figure 1. Figure 3 shows spectra of the three spot measurements performed using the 1.2 mm beam.

When analyzing the black writing, the main components are found to be iron and calcium, where the Ca is believed to be arising from the paper. Smaller amounts of Cu that could be attributed either to contamination as inks used to be stored in copper pots or could in some cases be attributed to the green color on the back side of the manuscript. The trace amounts of Pb and Hg are also attributed to contamination.

Figure 3 (b) is a scan of the red writing. The figure shows that the main element in this pigment is Pb with small amount of other elements due to reasons similar to those cited above for the black ink pigment. The green ink on the other hand (figure 3 (c)) shows the existence of two major elements which are Cu and Pb. (in addition to the minor elements arising from the paper and contaminations). It is obvious that the main component of the green pigment scan is Cu and lead is most probably due to fluorescence of the red pigment on the backside of the page.

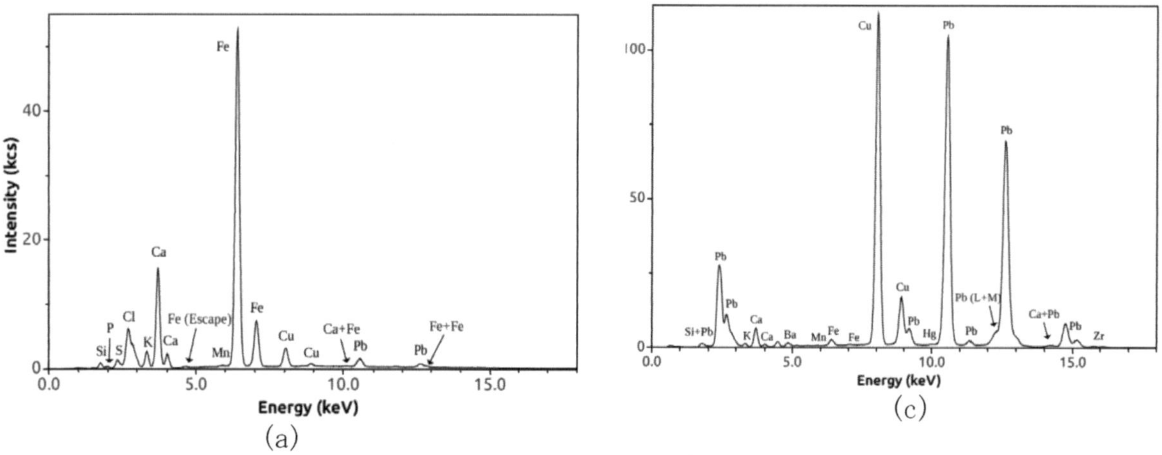

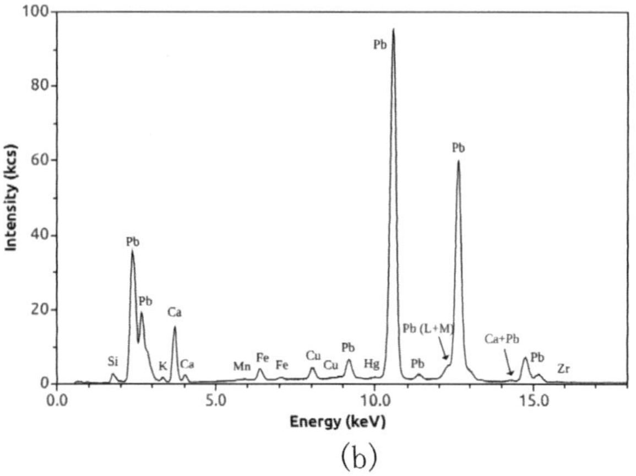

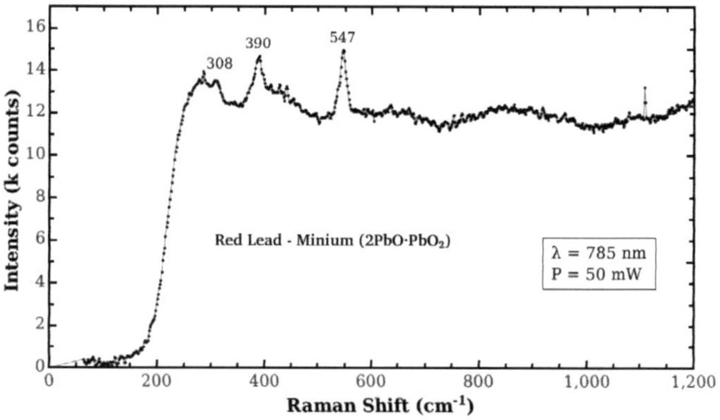

Figure 3: XRF spectra for spot measurements of (a) black, (b) red and (c) green color.

Figure 4: RAMAN spectra of red color pigment (red lead)

Figure 4 shows the RAMAN spectra of the red pigment used in this manuscript. The spectrum shows peaks of RAMAN shifts at (308, 390 and 547) cm^{-1}. These shifts are consistent with the shifts arising from “red lead" color [3]. The results of RAMAN spectr combined with those of XRF (mapping and spot analysis) confirm that the red pigment used in this manuscript is “red lead” known as minium (Pb_3O_4 sometimes written as $2PbO \cdot PbO_2$). Similar RAMAN measurements were performed on the black and green inks showed that the Cu-based green pigment similar to those used in old Islamic manuscripts is malachite ($CuCO_3$. $CuO(H)_2$)[3-6], while the black color is iron gall ink consisting of vitriol (FeSO4), tannic acid (from gall nuts),Arabic gum and water [5]. Different pigments were commonly used in the period 15th -16th century for black ink, namely, soot ink and blister**,** carbon-based ink prepared from wood-ash, sepia; an animal product from squid and cuttlefish or iron gall. Our measurements confirm

IOP Conf. Series: Materials Science and Engineering **37** (2012) 012006 doi:10.1088/1757-899X/37/1/012006

that iron gall was used as a black ink pigment in this manuscript [3-6]. These results are consistent with measurements performed by Van Der Snickt et al., on an illuminated sheet of parchment from a fifteenth or sixteenth century book of tides where iron gall was used for black ink pigment and green the Cu-based malachite was used for the green ink pigment. On the other hand, the red color in our manuscript was red lead while van der Snickt et al. found that HgS (cinnabar) was used for the red ink in their old manuscript [5]. The red lead pigment observed in the present manuscript was also observed in a Thai manuscript by Ermin et al [7]. The use of malachite for green ink and both red lead and cinnabar was common in pigments in old manuscripts. For example, Sawczak et al., found these pigments in wall paintings located in the Little Christopher chamber of the Main Town Hall in Gdan´ sk, Poland discovered in 1891 [8]

3.2 Manuscript 2

The second document is a fragment from a manuscript known as "Fatwas of Qari' Al-Hidaya". Figure 5 shows scans for the two sides of the fragment. The μ-XRF mapping area is shown in figure 5(c).

(a) (b) (c)

Figure 5: (a,b) The two sides of manuscript 2: and (c) mapping area of side 2

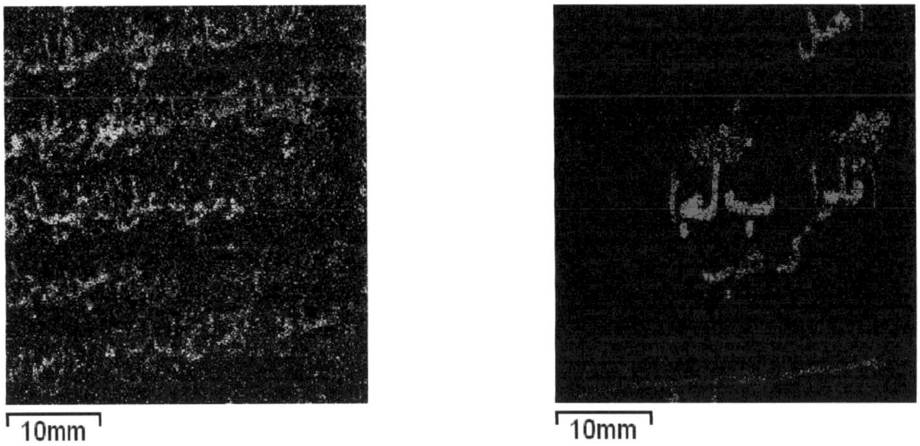

(a) (b)

Figure (6): μ-XRF Elemental maps of (a): Fe and (b) for Hg

Figures 6 shows μ-XRF elemental maps of side 2 of manuscript 2 obtained using the 10 μm beam and 50 kV energy . The maps show that black color is Fe-based gall ink pigment while the red color is Hg based HgS known as Vermillion- Cinnabar pigment. Cinnabar was also a common pigment used for red color after the 16th century [1, 2-4].

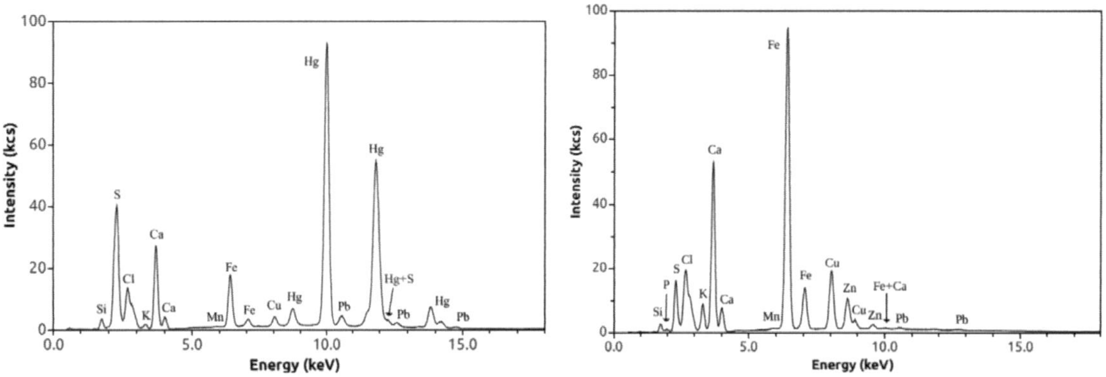

Figure 7: XRF spectra of red and black colors of manuscript 2.

Figure 7 shows XRF scans for the red and black colors respectively. As in the previous scans (Figure 3), the 1.2 mm x-ray beam was used with energy of 50 keV. The main peaks of figure 7 (a) are identified as Hg and S with Fe, Ca and others as trace elements coming from the paper. Small amounts of Pb were also observed as impurity associated with the Hg. The XRF results provide a strong indication that the red pigment in this manuscript is basically HgS (Cinnabar) [3-5]. Figure 7 (b) shows the XRF scan for the black writing. The figure shows that the main component of this pigment is Fe, while other lines in the spectra are coming from the paper (Ca) and impurities (Cu from ink storage containers), traces of Pb, K, S and Zn. These results also provide a strong indication that the black color ink is iron gall. These results are consistent with published results obtained for medieval manuscripts [5-8].

3.3 Manuscript 3

The third manuscript used in this investigation is shown in figure 8. XRF spectra for both black and red color are shown in figure 9. The black pigment shown in Figure 9 (a) is clearly an iron-based pigment with elements coming from the paper and trace elements mainly due to contamination and a strong Hg peak that is arising from the red color on the backside of the page. This indicates that the black pigment is iron gall ink [2-8].

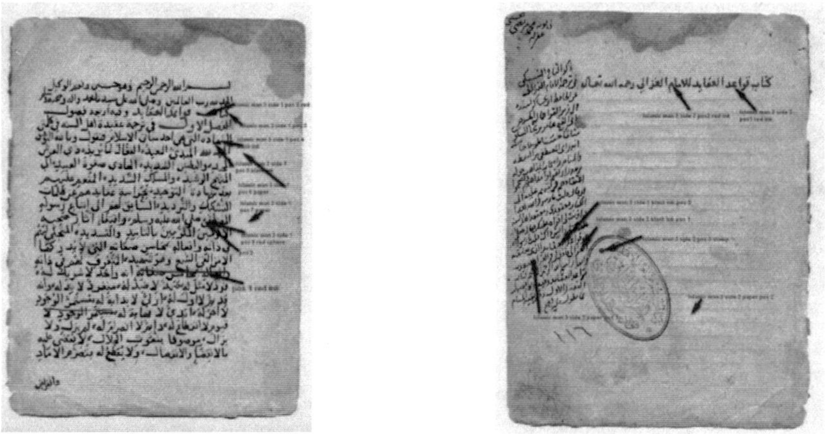

Figure 8: The two sides of a fragment of manuscript 3:

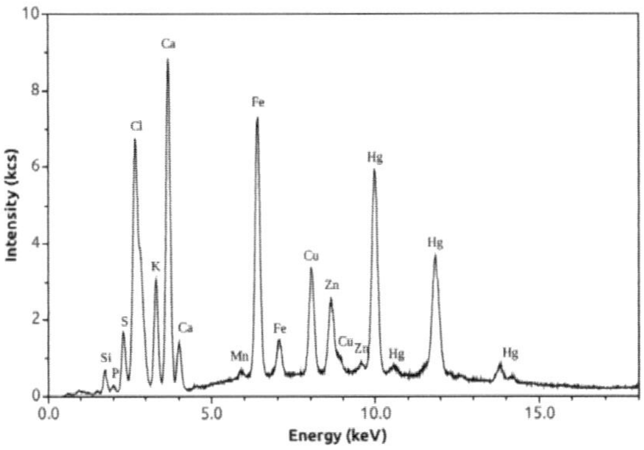

(a)

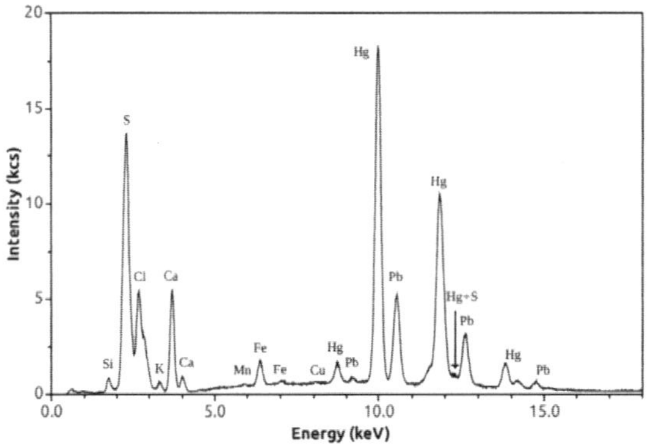

(b)

Figure 9: XRF Spectra of the (a) black and (b) red colors in manuscript 3.

Spectra were also obtained for different red spots. Figure 9 (b) shows a sample of these spots. The red color ink seems to be a mixture of both Hg-based and Pb-based pigments, which is unusual for Islamic manuscripts of this period. This result was also confirmed by elemental maps that are not shown in this work. We exclude the possibility of contamination because this result was consistent in all the scans on various red colored letters of the manuscript with almost the same Hg/Pb ratio, the Hg-based pigment being the major component. One possibility is that the red color has been overwritten in a later time or it was simply a mixture of cinnabar (HgS) and red lead (Pb_30_4 sometimes written as, $2PbO \cdot PbO_2$). We are currently carrying further investigations on this manuscript that will include high resolution elemental mapping, and RAMAN spectroscopy particularly on the red color. Vermillion- Cinnabar- HgS: was first use around the 16th century [9]. The presence of HgS as a major compound in the red color of this document is either an indication of forgery or an indication that the current fragment was written a few centuries after Al-Ghazalie's death. Al-Ghazali has written many books during his life only 72 of which are confirmed to be written by him as indicated by a letter he wrote to the Sultan Sanjar before he died. In that letter, he mentions that he wrote more than 70 books. Scholars classified some 457 books attributed to Al-Ghazali [10].

4. Conclusions:

The integration of μ-XRF mapping and spot analysis with μ-RAMAN spectroscopy provide complementary information about compounds used in various color pigments in ancient manuscripts. This was demonstrated on three Islamic manuscripts where various compounds used in these documents were identified. The iron gall pigment was used for the black ink in the three manuscript fragments investigated in this study, while malachite was the pigment used for green ink in the first manuscript. Red lead (Pb_30_4) was used for the red color ink in the first manuscript, while cinnabar (HgS) was used in the second manuscript. The red color in the third manuscript was a mixture of both red lead and cinnabar. The presence of cinnabar in a manuscript that was believed to be written in the eleventh century raises doubts about the authenticity of this manuscript. Information on pigments used in manuscripts will be useful for restoration and conservation scientists and provide useful information about materials used during various periods and in different regions of the Middle East. It will also provide information about authenticity and the age of these manuscripts.

Acknowledgment: we would like to thank Mr. Mohammed Nidal Kallah, from Juma Al Majid Center for Culture and Heritage in Dubai for providing the fragments and for useful discussions. NMH acknowledges the support of AUS through a Faculty Research Grant.

IOP Conf. Series: Materials Science and Engineering **37** (2012) 012006 doi:10.1088/1757-899X/37/1/012006

References

[1] A. Duran, J. L. Perez-Rodriguez, T. Espejo, M. L. Franquelo, J. Castaing ,P. Walter, Anal Bioanal Chem (2009) **395**:1997–2004.
[2] L. Burgio, R. J. H. Clark, and R. R. Hark, PNAS, March 30, 2010, vol. **107**, no. 13, 5726–5731
[3] M. Aceto, A. Agostino, E. Boccaleri, F. Crivello, J. Raman Spectrosc. 2010, **41**, 1434–1440
[4] L. Burgio, R. J. H. Clark, R. R. Hark, M. S. Rumsey, and CA. Zannini, Applied Spectroscopy, **63**, Number 6, 2009, 611-620.
[5] G. Van Der Snickt, W. De Nolf, B. Vekemans, K. Janssens, Appl. Phys. **A 92**, 59–68 (2008).
[6] Nabil Nofal, Prospects: the quarterly review of comparative education, (Paris, UNESCO: International Bureau of Education), vol. XXIII, no. 3/4, 1993, p. 519-542.
[7] K. Eremin, J. Stenger, Jo-Fan Huang, A. Aspuru-Guzik, T. Betley, L. Vogt, I. Kassal, S. Speakman and N. Khandekar; J. Raman Spectrosc. 2008; **39**: 1057–1065.
[8] M. Sawczak, A. Kamin´ska, G. Rabczuk, M. Ferretti, R. Jendrzejewski, G. S´liwin´ski; Applied Surface Science 255 (2009) 5542–5545
[9] Gettens, R. J., Feller, R. L. & Chase, W. T., *Artists' Pigments: A Handbook of their History and Characteristics*, New York: Oxford University Press, 1993, p. 159
[10] Al Ghazali's Works, Cairo, 1981, Abdul-Rahman Badawi

IOP Conf. Series: Materials Science and Engineering **37** (2012) 012007 doi:10.1088/1757-899X/37/1/012007

Results of the Italian neu_ART project

A Re[1], F Albertin[1], C Bortolin[2], R Brancaccio[1], P Buscaglia[3], J Corsi[2], G Cotto[1,2], G Dughera[1], E Durisi[1,2], W Ferrarese[1,2], M Gambaccini[4], A Giovagnoli[3], N Grassi[3], A Lo Giudice[1,2], P Mereu[1], G Mila[1,2], M Nervo[3], N Pastrone[1], F Petrucci[4], F Prino[1], L Ramello[5], M Ravera[3], C Ricci[1], A Romero[1,2], R Sacchi[1,2], A Staiano[1], L Visca[1,2] and L Zamprotta[1,2]

[1] Istituto Nazionale di Fisica Nucleare, Sezione di Torino, via Pietro Giuria 1, 10125 Torino, Italy
[2] Dipartimento di Fisica Sperimentale, Università di Torino, via Pietro Giuria 1, 10125 Torino, Italy
[3] Centro Conservazione e Restauro "La Venaria Reale", Piazza della Repubblica, 10078 Venaria Reale, Torino, Italy
[4] Dipartimento di Fisica, Università di Ferrara and Istituto Nazionale di Fisica Nucleare, Sezione di Ferrara, Via Saragat 1, 44100 Ferrara, Italy
[5] Dipartimento di Scienze e Tecnologie Avanzate, Università del Piemonte Orientale, Viale Teresa Michel 11, 15121 Alessandria, Italy

Email: alessandro.re@to.infn.it

Abstract. The neu_ART project aims at developing *state of the art* transmission imaging and computed tomography techniques, applied to art objects, by using neutrons as well as more conventional X-rays. In this paper a facility for digital X-ray radiography of large area paintings on canvas or wooden panels and for the X-ray tomography of large size wooden artifacts, recently installed in a protected area, is presented. The results of a K-edge radiography facility that will soon be installed in the same area are also shown.

1. The neu_ART project

The neu_ART project ("neutron and x-ray tomography and imaging for cultural heritage"), financed by the Piedmont region (Italy), is a collaboration among the University of Torino, the Istituto Nazionale di Fisica Nucleare (INFN) and the restoration centre CCR "La Venaria Reale". Among the goals of this project is the construction of an integrated X-ray imaging system for computed tomography (CT) and digital radiography (DR) of big artworks and of a facility for K-edge radiographies of small portions of paintings. The benefits of such imaging techniques in the field of cultural heritage are widely acknowledged [1,2]. Their application is useful as preliminary step for a restoration intervention, giving significant data about conservation, history and structural technique.

2. X-ray digital radiography

The instrument provides high quality radiographic images of paintings up to 3×4 m^2 by scanning horizontal slices of their surface with an X-ray linear detector. The painting is positioned vertically in front of a motorized mechanical system allowing to select the height to be scanned. A horizontal axis, with an accurate speed control, synchronizes the detector movement with the image acquisition software. The speed can reach values up to 6.5 m/min, depending on the intensity of the signal.

The X-ray source, used both for DR and CT, is a General Electric Eresco 42MF4. It has a Tungsten anode with a focal spot size of 1.5 mm and a maximum current of 4.5 mA at the upper voltage of 200 kV, increasing to 10 mA at lower operating voltages. The detector for radiography is a Hamamatsu C10650: it's a linear TDI (Time Delay Integration) CCD detector, coupled with a fiber optic plate and a scintillator. It features a CCD 22 cm long with a pixel size of 48 μm for a total pixel number of 4608×128 and is designed for operations with X-ray energies lower than to 95 keV.

The instrument was tested on a painted wooden panel (dimensions: 30×41.5×3.5 cm^3) realized by the CCR (figure 1a): it's a copy of a Gentile da Fabriano where materials and techniques match those used by the author. The X-ray radiography (figure 1b) was obtained with a tube voltage of 80 kV and a current of 7 mA by placing the tube and the panel at a distance of about 160 cm and 2 cm from the detector respectively. The detector scanning speed was 1 m/min. The aim of this radiography was to check how many information useful to a restorer could be obtained to evaluate the condition of conservation of the object. From the radiography several important features can be observed: the painting was realized on two tables joined together by a radiopaque glue (the bright line in the center of the figure) and by three wooden pins of different shapes (figure 1e); the construction technique of the frame (figure 1c); wooden dowels were placed instead of a knag (figure 1d). Note that despite the high X-ray energy, necessary to transverse the thickness of the panel, prevents a good contrast on thin layers, the pigment layer is clearly visible in some parts of the painting (figure 1b), especially where Pb white is used for the flesh tones.

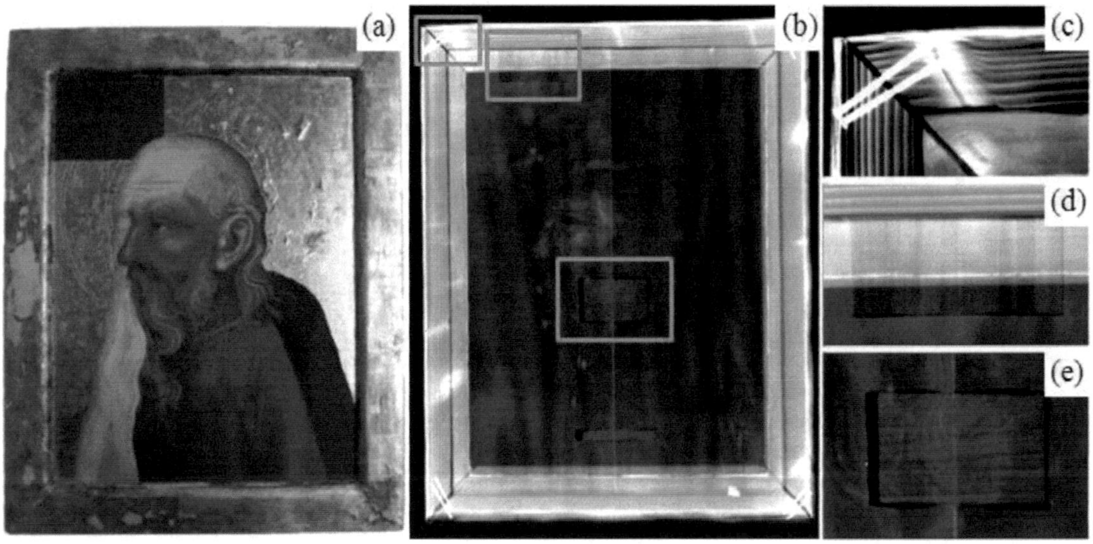

Figure 1. Copy of a wooden painted table by Gentile da Fabriano (a) compared to the X-ray radiography (b) with some enlarged details: nails in the frame (c); wooden dowels (d); wooden pin (e).

3. X-ray computed tomography

The CT is conveniently used in many of the restoration activities at CCR, especially where wooden sculptures, furniture, paintings on panel and ceramics are involved. It provides more information with respect to DR, especially for those objects in which the three dimensional information is important. It allows studying the executive and assembly techniques for both wooden sculptures and furnishing without actual dismounting, providing the possibility to discriminate the presence and the location of different materials, as different wooden species, and to evaluate thickness and dimensions of the various parts.

The mechanical system and the X-ray tube used for the CT are the same described in section 2 with the addition of a rotary stage of 160 cm diameter to rotate large objects in small steps with high precision. The dimensions of the objects to be analyzed require using high X-rays energies. To override the energy limitations of the detector used in DR (see section 2), a different detector able to operate up to the maximum tube energy of 200 keV was selected: a Hamamatsu C9750-20TCN linear detector with 2560 pixels of 200×200 μm^2 each, with a total length of 51.2 cm.

The setup was tested on a sample prepared to simulate typical wooden widths of an empty statue or furniture: it's a piece of poplar (figure 2a) with a maximum width of about 25 cm and four steps of about 5 cm each. The CT was obtained operating the tube at a voltage of 180 kV and a current of 5 mA, placing the tube and the center of the rotary stage at distances of 295 cm and 80 cm from the detector respectively. The detector speed was of 5 m/min and 720 radiographies were acquired, one each 0.5°. The CT reconstruction [3] was performed with a non-commercial software-utility developed by Dan Schneberk of Lawrence Livermore National Laboratory (USA). One of the radiographies is shown in figure 2b: it reveals some features of the wood, but it's impossible to determine at which depth they are. In the CT reconstruction (figure 2c) many more details as tree-rings, knags, cracks are clearly visible and their position can be precisely determined.

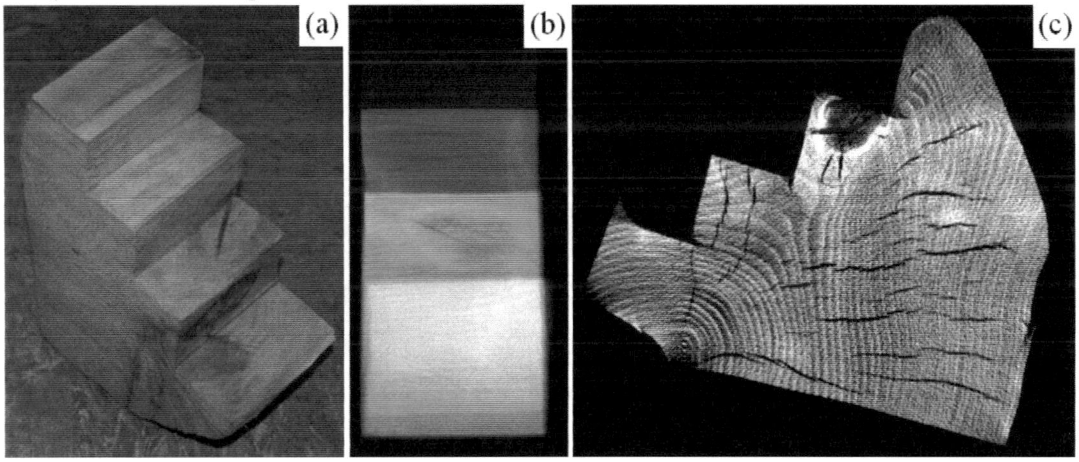

Figure 2. A wooden step sample (a), X-ray radiography (b) and tomographic reconstruction (c).

4. K-edge radiography

The K-edge technique [4-8] takes advantage of the sharp rise of X-ray absorption coefficient of an element, the K-edge discontinuity, to map it on the whole surface of the painting. Each pigment is characterized by one or more elements; mapping an element means finding the spatial distribution of that pigment. The procedure requires two quasi-monochromatic X-ray images (in the energy range of 8-40 keV) above and below the K-edge energy of the element under investigation. This leads to a significant variation of the target element signal, maintaining almost unchanged the response from the background. A quasi-monochromatic X-ray beam can be achieved via Bragg diffraction of a standard X-ray beam on a mosaic crystal [9]. The images are processed by a subtraction algorithm [4] to obtain two new images: one gives the mass density distribution (g/cm^2) of the element under study and the other gives the distribution of all other materials in the sample.

The results presented in this paper were obtained using an X-ray source, installed in the LARIX Laboratory of Ferrara Physics Department. A SSD detector was used to perform X-ray imaging [10,11]: it consists of 512 Si-strips, 1 cm thick and 300×100 μm pixel size. It is an edge-on detector and it works in single photon counting mode. The edge-on configuration allows a larger penetration thickness to increase efficiency in the 20-40 keV energy range. To ensure a very small dead area, the detector is cut perpendicular to the strips at only 20 μm from the end of the strips. Counting measurements were carried out by comparison with a CZT detector to compute the SSD efficiency which resulted (81±4)% at 9 keV and (88.3±1.4)% at 26.7 keV.

Elemental distributions on test objects and canvas with different pigment layers were achieved. At lowest energy, elemental mapping of Cobalt (7.7 keV K-edge energy) and Copper (8.97 keV) on Smaltino (potassium glass containing cobalt) and Azurite ($2CuCO_3 \cdot Cu(OH)_2$) painting layers was performed (figure 3a). According to the Lehmann algorithm, the element distribution (g/cm^2) on the analyzed surfaces is presented using a grey level scale: higher grey levels correspond to a greater quantity of the element under study; in the same figure the average Co and Cu contents in each pixel column are shown. The diagnostic potential of this technique to separate and map different elements is shown in figure 3b, where the elemental mapping of Antimony (30.49 keV K-edge energy) and Cadmium (26.71 keV) on superimposed layers of Naples Yellow ($Pb_2Sb_2O_7$) and Cadmium Red (CdS) is presented.

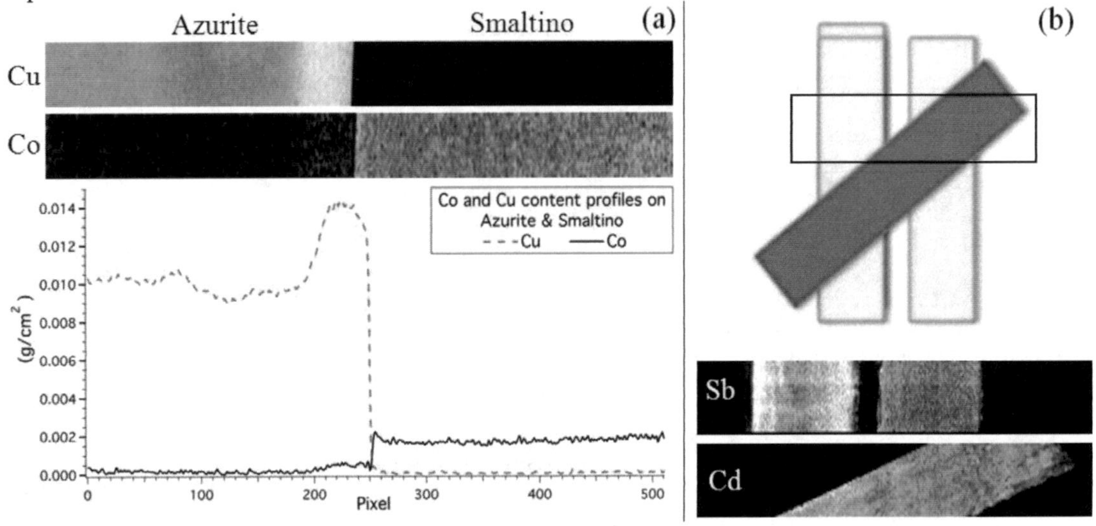

Figure 3. (a) images of Cobalt and Copper distribution detected in the Azurite (left) and Smaltino (right) target objects; in the graphic the average Co and Cu content measured on images is shown; (b) target object structure: a single Cadmium Red pigment layer (diagonal); double (on the left) and single (on the right) Naples Yellow layers pigments (vertical). Sb and Cd detected in the imaged frame.

References

[1] Casali F 2006 *Physical Techniques in the Study of Art, Archaeology and Cultural Heritage* vol 1 (New York: Elsevier) chapter 2 pp 41-123
[2] Bettuzzi M et al 2004 *Physics Methods in Archaeometry* (Amsterdam: IOS Press) pp 461-470
[3] Kak A C and Slaney M 1988 *Principles of Computerized Tomographic Imaging* (Piscataway, NJ: IEEE Press)
[4] Lehmann L A, Alvarez R E, Macovski A, Brody W R 1981 *Med. Phys.* **8**(5) 659
[5] Lewis R 1997 *Phys. Med. Biol.* **42** 1213-43
[6] Baldelli P, Taibi A, Tuffanelli A, Gilardoni M C, Gambaccini M 2005 *Phys. Med. Biol.* **50** 2225
[7] Baldelli P et al 2006 *Il Nuovo Cimento* **29C**(6) 663-72
[8] Krug K et al 2006 *Appl. Phys. A* **83** 247
[9] Albertin F et al 2009 *Appl. Phys. A* **96**(2) 503-10
[10] Bolanos L et al 2009 *Nucl. Instrum. Methods A* **608** 410-6
[11] Albertin F 2011 *Il Nuovo Cimento* **34**(5) 227-35

Acknowledgments

This study was carried out in the framework of the "neu_ART" research project funded by Regione Piemonte. We would like to thank the valuable work of the team in the Technological Laboratory of INFN Torino, in particular S. Brasolin, F. Borotto, G. Ferrero and R. Panero. We thank S. Chiozzi, S. Squerzanti (INFN Ferrara) and L. Landi (Ferrara University) for the construction and development of the K-edge facility in Ferrara and F. Evangelisti (INFN Ferrara) who is also contributing to the new K-edge facility design. Thanks to E. Ghio for studying the main characteristics of Gentile's technique.

IOP Conf. Series: Materials Science and Engineering **37** (2012) 012008 doi:10.1088/1757-899X/37/1/012008

A newly developed, portable, vacuum-chamber equipped XRF-instrument, designed for the sophisticated needs of the Kunsthistorisches Museum, Vienna

K Uhlir[1], B Frühmann[1], G Buzanich[2,3], M Griesser[1], C Streli[2], P Wobrauschek[2], B Grossmayer[2], S Smolek[2]

[1] Conservation Science Department, Kunsthistorisches Museum, Burgring 5, A-1010 Vienna, Austria
[2] Atomic Institute (ATI) – Vienna University of Technology, Stadionallee 2, A-1020 Vienna, Austria
[3] BAM Federal Institute for Materials Research and Testing, D-12200 Berlin, Germany
E-mail: katharina.uhlir@khm.at

Abstract

In the analysis of artworks x-ray fluorescence (XRF) spectroscopy gains a high significance because of the non-destructiveness of the method. It can provide valuable information of technologies and materials and support conservation and restoration efforts. This method is especially suitable for the study of the collections of the Kunsthistorisches Museum Vienna (KHM) with its affiliated institutions, because of the great diversity of the materials and objects with different art historical and ethnological background. Within a research project in co-operation of the Conservation Science Department of the KHM with the ATI (Atomic Institute of the Vienna University of Technology) and the IAEA (International Atomic Energy Agency) a portable XRF instrument ("PART II") was designed and constructed, optimized to cope with the diverse analytical problems of a wide-spread museum collection as it is kept at the KHM.
Beneath some technical improvements concerning the optimization of the design of the new instrument (regarding stability and accessibility to objects) with respect to its precursor model, a special focus was laid on the detection of light elements to be able to analyze chemical elements from sodium (Na) onwards. In this way the palette of materials that can be analyzed could be expanded to matters like glass and enamel. In this paper the advantages of the new facility will be discussed.

Introduction

For the investigation of museum objects it is very important to apply a non-destructive analytical method. Additionally, as many objects are not moveable or too valuable for transportation to an external research facility, the investigations should be, as far as possible, done at the museum itself. Therefore, the instrument has to be portable to be applied in-situ in the exhibition spaces. Especially for the Kunsthistorisches Museum (KHM with its affiliated institutions: MVK – Museum of Ethnology, ÖTM – Austrian Theatre Museum, Treasury, Neue Burg with Ephesos Museum - Collection of Ancient Musical Instruments - Collection of Arms and Armour, Museum of Carriages and Department of Court Uniforms and Ambras Castle), which holds many objects of different materials and different art historical background, problems of a great diversity have to be considered. An analytical method which is employed very successfully for the investigation of cultural heritage objects and that can fulfill all the above mentioned needs, is the energy-dispersive x-ray fluorescence (XRF) analysis. Valuable information about materials and pigments used in art and ethnological objects, which are of great interest for restoration and conservation purposes, can be obtained using XRF. Additionally, an integral description of the objects in an art historical and ethnological background is possible. For these reasons also the KHM was interested in the application of this technique in its institutions.
A prototype XRF instrument (PART I – *P*ortable *Art* Analyzer I) [1, 2] developed by the IAEA (International Atomic Energy Agency, Laboratories Seibersdorf) and the ATI (Atomic Institute of the Vienna University of Technology) could be applied and tested at the museum [3]. After this testing phase some ideas for an improvement of the instrument were born [4], which could be put into practice by the construction of an enhanced XRF instrument, funded by the Austrian Science Fund (FWF, project no. L430-N19; „Portable Art Analyzer – PART").

Experimental

The goal of the mechanical design of the new portable focused beam XRF spectrometer was to build a transportable system with improvements in the accessibility to the different parts of the objects and in the positioning of the instrument in front of an object as well as an enhancement in the detection of low-Z elements.

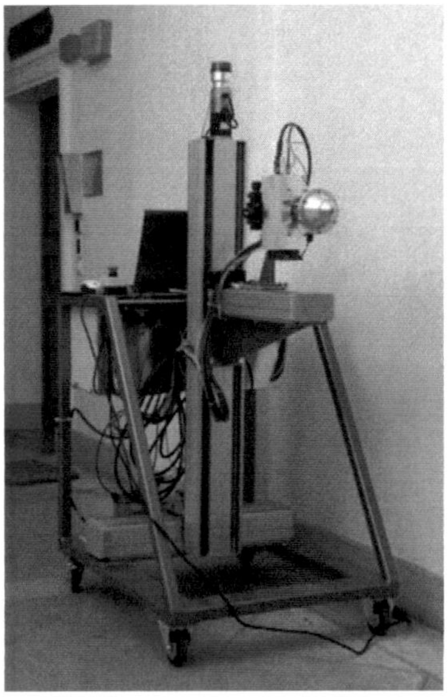

Fig. 1 Portable Art Analyzer II (PART II)

The resulting instrument was called PART II (fig. 1). The whole instrument is mounted on a stable frame with steerable and lockable wheels for transportation. The measuring head is mounted on three electronically controllable translation stages in x-, y- and z-direction. In the z-direction (up and down) the traverse amounts 100 cm, in x- and y-direction 30 cm, respectively. These large traverse distances are necessary to be able to analyze also big objects or paintings. The new design of a stable stand combined with the ability of moving the measuring head very flexible (allowing an overhang up to 40 cm in the forefront of the frame) enables measurements with fewer restrictions to the size and shape of the objects and their holders. For example, possible collisions between the stand of the instrument and the stand of an easel where the paintings are mounted can be avoided.

The laptop is placed on a countertop directly on the frame, so that all the equipment necessary for performing measurements is positioned close together. All electronic components are mounted on a separate panel under the laptop-countertop.

One of the most striking construction highlights of PART II is the measuring head with the vacuum chamber (fig. 2). It is mounted rotatable (horizontal) for better accessibility to the objects and can be locked in the desired position. The design of the vacuum chamber is based on the need of good accessibility to the different parts of the museum objects. Therefore, a conical shape with a tip was employed. Both, incident and fluorescent beams pass through a Kapton™ window of 8 μm thickness, which is applied at the tip of the vacuum chamber. In contrast to PART I [1, 2], the Kapton™ window is pre-stressed and glued directly on the outside of the chamber, which leads to less bending of the window when the vacuum is applied. The air gap between chamber and sample is about 1 mm, minimizing absorption losses in the excitation and x-ray fluorescence radiation paths and enhancing the sensitivity for light elements.

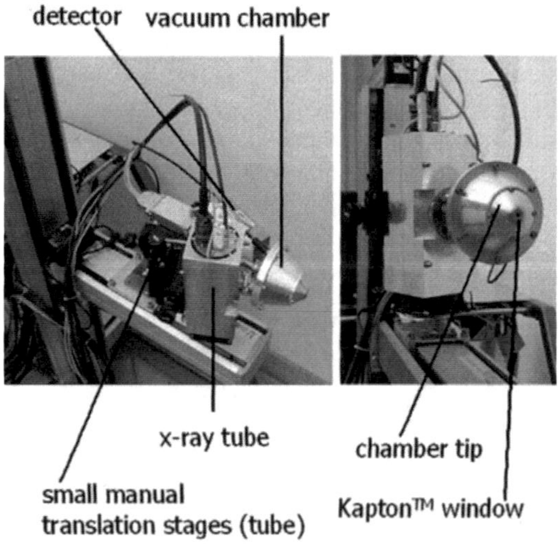

Fig. 2 The measuring head of PART II

The internal design of the vacuum chamber [4] is shown in fig. 3. The x-ray optic used is a polycapillary (focal spot size 145 μm for Cr K_{α}, 5.41 keV). The detector is a Si Drift Detector (SDD) with an improved efficiency in comparison to PART I [1, 2, 4]. The alignment of the incidence and take-off angles of radiation allows an acute-angled design of the vacuum chamber [4]. Inside also a camera for the positioning of the instrument in front of an object and an LED for the illumination is incorporated [4]. Additionally, two collimated laser pointers cross in 1 mm distance outside the Kapton™ window (focal point of the capillary) [4].

IOP Conf. Series: Materials Science and Engineering **37** (2012) 012008 doi:10.1088/1757-899X/37/1/012008

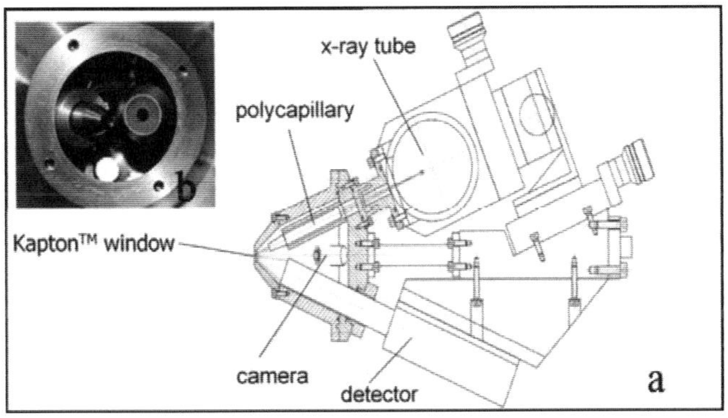

Fig. 3 Sketch of the internal design of the vacuum chamber (a) and a frontal look inside the chamber (b)

For the operation of the equipment a Mo- and a Cr-x-ray tube (point focus of 200 μm, maximum power of 50 W) of Oxford instruments [4] are available. The change between the tubes can be done manually in a fast and very simple manner. The Mo-tube gives a good overview of all components that can be expected in the museum collection. Nevertheless, the L-lines of Mo coincide with the sulfur and the chlorine lines, both components of historic glass and enamel. To obtain an undisturbed spectrum of the low-Z elements the Cr-tube was acquired.

Using PART II with the Mo-tube the detection limits could be raised compared to PART I (equipped with a Pd-tube) [4]. The analysis of the NIST glass standard reference material 620, measured using both instruments (measuring conditions: 50 kV, 1 mA, 1000 s), shows an overall enhancement of intensity with PART II, doubling the intensity of the Na-peak from 914 to 2010 counts (fig. 4). For comparison all peak intensities of the two spectra are listed in tab. 1. For fitting the WinQXAS program of the IAEA was used.

Using PART II with the Cr-tube, due to an overload of the detector, no spectrum with the conditions 50 kV and 1 mA could be obtained. Therefore, the values of a spectrum obtained with 50 kV, 0.6 mA and 1000 s (fig. 5) were extrapolated for comparison (tab. 1).

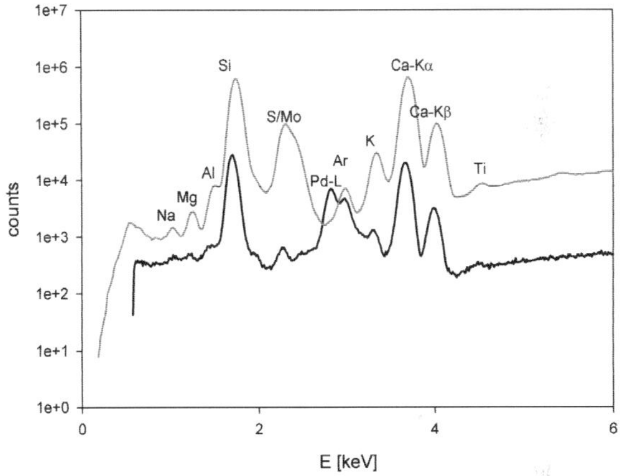

Fig. 4 Two spectra of NIST SRM 620 collected with PART I (black) and PART II with Mo-tube (grey). The energy-range up to 6 keV is shown. Measuring conditions for both spectra: 50 kV, 1 mA, 1000 s

oxide in NIST SRM 620	E (keV) $K_{\alpha 1}$	nominal concentration (oxides) NIST SRM 620	peak area [counts] PART I (Pd-tube)	PART II (Mo-tube)	PART II (Cr-tube)
Na_2O	1.041	14.39 %	914 ± 104	2010 ± 149	1380 ± 178
MgO	1.254	3.69 %	1858 ± 112	9206 ± 181	6270 ± 226
Al_2O_3	1.487	1.80 %	4864 ± 125	44395 ± 277	33415 ± 347
SiO_2	1.74	72.08 %	272897 ± 535	3869057 ± 1979	305277 ± 2293
SO_3	2.307	0.28 %	3912 ± 118	725827* ± 852	4087 ± 398
	2.957	Ar (air)	64660** ± 264	28966 ± 292	235950 ± 745
K_2O	3.313	0.41 %	10043 ± 160	191945 ± 519	857517 ± 1301
CaO	3.691	7.11 %	225817 ± 456	4390572 ± 1967	19658823 ± 5305
TiO_2	4.509	0.018 %	1620 ± 106	24944 ± 409	111550 ± 648
Fe_2O_3	6.399	0.043 %	11024 ± 169	151944 ± 683	15926 ± 357

Tab. 1 Comparison of the elemental peak intensities in the spectra of NIST SRM 620, collected with PART I and PART II. Measuring conditions for PART I and PART II with Mo-tube: 50 kV, 1 mA, 1000 s, PART II with Cr-tube: 50 kV, 0.6 mA, 1000 s extrapolated to 1 mA, * Mo L-lines subtracted, **Pd L-lines subtracted

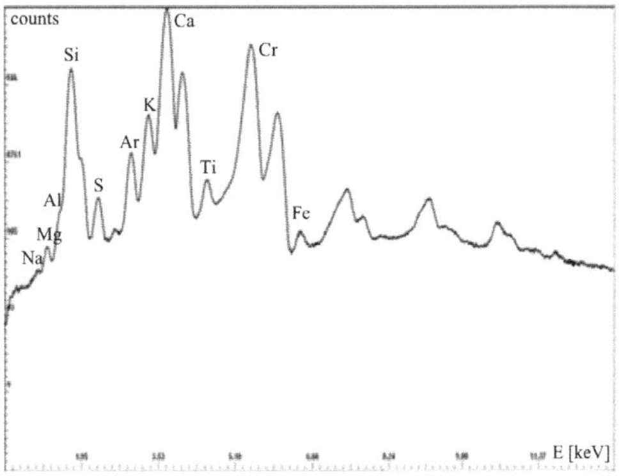

Fig. 5 Spectrum of NIST SRM 620 collected with PART II, Cr-tube (50 kV, 0.6 mA, 1000 s)

Comparing the peak areas in tab. 1, it turned out that the Mo-tube shows a better excitation of the K lines of Na – Si than the Cr-tube. This was not expected but can be explained by excitation with the Mo L-lines. Additionally, using the Cr-tube, pile-up peaks in the energy-range above Cr restrain a proper analysis of the element-lines in this area (fig. 5). For the elements K – Ti, on the other hand, the Cr-tube shows the best excitation. The choice of the proper tube will therefore strongly depend on the analyzed material and the accurate problem.

The detection limit for the PART II device using the Mo-tube could be calculated to 7.7 % Na [4], equaling 10.4 % Na_2O. Experimental, the three glass reference materials NIST 1831 (13.32 wt% Na_2O), NIST 620 (14.39 wt% Na_2O) and V5 (19.00 wt% Na_2O) were analyzed using the measuring conditions 50 kV, 1 mA, 500 s. The spectra were fitted with the WinAxil program and the counts of the Na-peaks are plotted versus the Na_2O concentration in fig. 6. For the lowest Na-content in NIST 1831 417 counts could be found (fig. 6). A Na-peak of 13 wt% Na_2O is detectable properly. Unfortunately, a reference material with approximately 10 wt% Na_2O was not available.

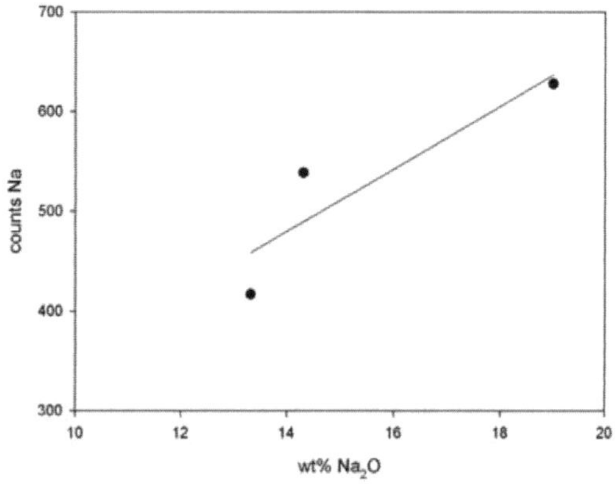

Fig. 6 Counts of the Na-peaks of the reference materials NIST 1831, NIST 620 and V5

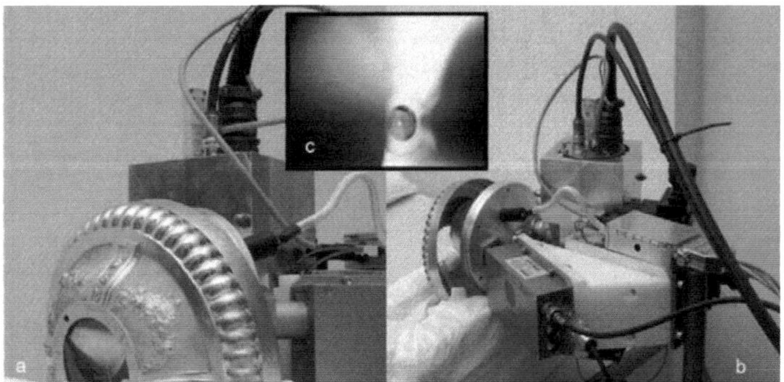

Fig. 7 The vacuum chamber of PART II positioned to measure in the inner part of the pedestal shown from the front (a) and the backside (b). The view of the camera through the Kapton™ window is also shown (c).

In fig. 7 an example for the potential of positioning is shown. The instrument is placed to analyze the gold-composition in the inner side of the pedestal of an Étagère, created in 1755 by Peter Josef Fonson. This object is made of red (or rose) gold, which was chemically etched to leach out the base metals copper and silver from the surface to achieve a golden appearance. Because XRF is a surface sensitive method, polishing of a small area is necessary to perform a quantitative investigation of the base alloy. Of course the polishing has to be done at a very inconspicuous part, like in the inner part of the pedestal.

Due to the improved design of the measurement head of the PART II instrument and even though the distance to the Kapton entrance window has to be 1 mm, measurements in the inner part of the object were possible. The delicate positioning is shown from the front and the backside in fig. 7. The view of the camera mounted inside the vacuum chamber looking through the Kapton window is also inserted in Figure 7. The two laser pointers cross in the focal point of the capillary (1 mm outside the chamber). Evaluation of the analysis, carried out using tht Mo-tube, 50 kV, 0.3 mA and 200 s measuring time, showed an alloy composition of approximately 70 wt% Au, 20 wt% Cu and 10 wt% Ag. One selected spectrum is presented in fig. 8, clearly showing the signals for Au, Ag and Cu. As expected, the analysis of metal alloys could be performed without major complications. Only between approximately 4 and 7.5 keV some diffraction peaks, due to the acute-angled design of the vacuum chamber, appear. For reducing these diffraction peaks (if they overlap with the peaks of important elements) an aluminum filter can be used.

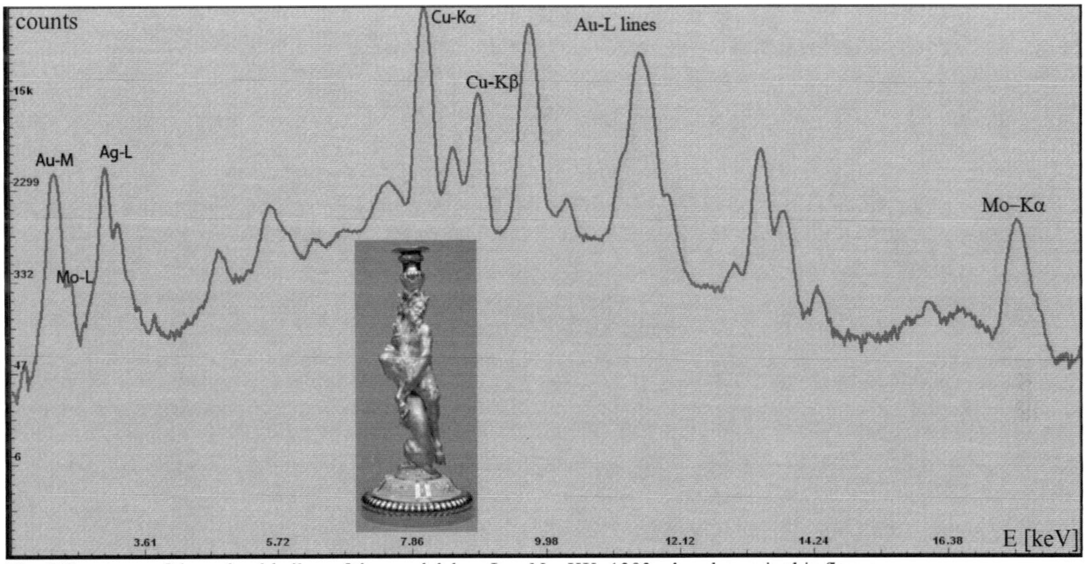

Fig. 8 Spectrum of the red gold alloy of the candelabra, Inv. No. KK_1283, also shown in this figure

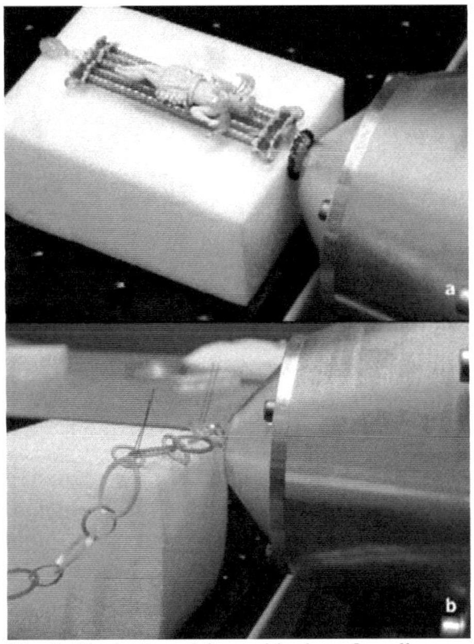

Fig. 9: Illustration of the positioning of small and very fragile glass objects in front of PART II: glass figurine Inv. No. KK_2801 (a) and glass chain Inv. No. KK_3042 (b).

One project concerning light elements was the analysis of 16th century glass jewelry (a unique glass jewelry collection held by the Collection of Sculpture and Decorative Arts, so-called "Kunstkammer", of the Kunsthistorisches Museum Vienna). Originally the pieces were part of Archduke Ferdinand's II "Kunst- und Wunderkammer", located at Ambras Castle near Innsbruck, Tyrol, Austria. For producing these outstanding objects he employed glass-makers from Murano, a very unique privilege, so the manufacturing technique should be based on Venetian recipes. In a first study around 80 objects made of differently colored transparent and opaque glasses were analysed. Representative for the whole collection two pieces will be presented here: a figurine (fig. 9a, Inv. No. KK_2801), where three differently colored glasses could be analyzed, and a glass chain (Fig. 9b, Inv. No. KK_3042). Positioning of these glass objects was often difficult due to their brittleness and different shapes. The mostly thin and highly varying thicknesses of the glass-parts as well as often unfavorable measurement geometries are the main problems for the quantitative evaluation of the data. Therefore, as a first step, only classifications of the glass types used were made.

The measurements were carried out using the Cr-tube (30 kV, 1 mA and 100 s measurement time per point, vacuum) to avoid any overlap in the low-energy region, taking into account that the identification of manganese (Mn) is problematic because of the overlap of Cr K_{β} and Mn K_{α}. Additionally, the presence of Na is not always assured, because of the necessity of applying short measuring times to get an overview of a vast number of objects within a very limited period of time available for analysis. Also corrosion plays a big part in the detectability of Na. Nevertheless, the K and Ca intensities can give an indication of the glass type used.

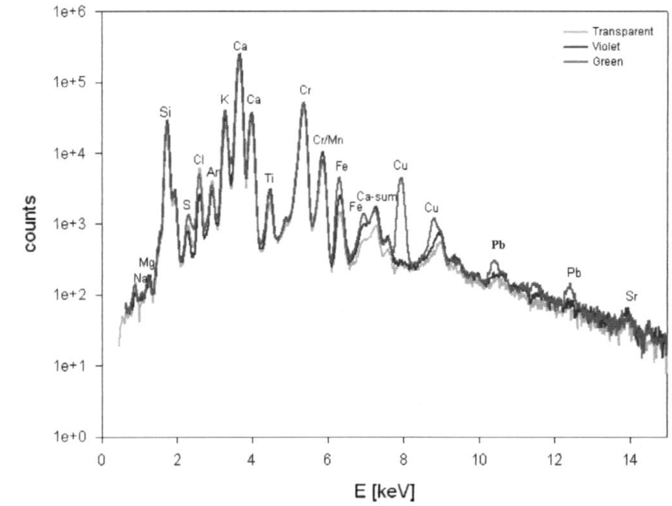

Fig. 10 Spectra for different areas on the figurine KK_2801 (see Fig. 9a) showing the characteristic elements for the specific colors.

The spectra of the differently colored glasses of the figurine KK_2801 are compared in fig. 10. As expected for these items, the three differently colored glasses show a similar composition for the base glass used. The applied technique of lamp working requires glasses with very similar physical characteristics. Therefore, the coloring elements were added to the same base glass - oxides of copper and iron for green, of manganese for violet. Soda-lime-silica glass could be identified for the base glass composition.

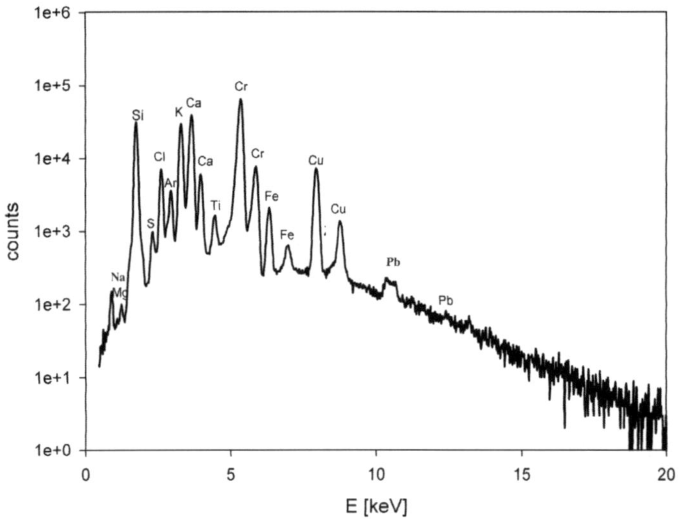

Fig. 11: Spectrum of the transparent blue glass of the glass chain KK_3042 (see Fig. 9b).

The spectrum of the blue glass of the glass chain KK_3042 is displayed in fig. 11. It can be seen that the Ca-concentration is very low compared to the spectra in fig. 10. These low concentrations of stabilizer can be found in the so-called Venetian *cristallo* glass, which was obviously used for some blue hues in the analyzed items. This seems also to be the reason why these glasses are more affected by corrosion. The coloring in this blue part is due to copper oxide.

IOP Conf. Series: Materials Science and Engineering **37** (2012) 012008 doi:10.1088/1757-899X/37/1/012008

Conclusion

In summary it can be stated that the portable XRF spectrometer ("PART II") was successfully designed to meet the needs that arise within the miscellaneous collections of the KHM: optimization of the vacuum chamber (minimization in size and conical form), minimization of the pathways of the excitation and fluorescence beam in air due to a better fixation of the Kapton™ window of the vacuum chamber, high efficiency of the detector and optimization of the positioning of the measurement head by the use of three electronically controllable translation stages. As could be shown, the qualitative analysis of elements down to Na is definitely possible, although the Cr-tube, acquired with regard of a better excitation of the light elements, did not fulfill the expectations. Nevertheless, following the first measurements on standards, it is very likely that with the Mo-tube semi-quantitative studies of glass and enameled objects will be possible using the PART II.

Acknowledgements

The authors gratefully thank A. Markowicz and his team from the International Atomic Energy Agency (IAEA), Seibersdorf Laboratories, for their invaluable support in constructing the PART instrument and the Austrian Science Fund (FWF) for financing the project "*P*ortable *ART* Analyser (PART)", project no. L430-N19.
Special thanks to Eva Putzgruber for the excellent collaboration regarding the analysis of the 16th century glass jewelry.

References

[1] Buzanich G, Wobrauschek P, Streli C, Markowicz A, Wegrzynek D, Chinea-Cano E, Bamford S 2007 A portable micro-XRF-Spectrometer with polycapillary optics and vacuum chamber for archeometric and other applications *Specrochimica Acta* B **62/11** pp 1252-1256
[2] Buzanich G 2008 *Portables Röntgenfluoreszenzspektrometer mit Vakuumkammer: In situ Analysen von Objekten aus Kunst und Archäometrie* (VDM Verlag Saarbrücken)
[3] Uhlir K, Griesser M, Buzanich G, Wobrauschek P, Streli C, Wegrzynek D, Markowicz A, Chinea-Cano E 2008 Applications of a new portable (micro) XRF instrument having low-Z elements determination capability in the field of works of art *X-ray spectrometry* **37/4** pp 450-457
[4] Buzanich G, Wobrauschek P, Streli C, Markowicz A, Wegrzynek D, Chinea-Cano E, Griesser M, Uhlir K 2010 PART II (Portable ART analyzer) development of a XRF spectrometer adapted for the study of artworks in the Kunsthistorisches Museum, Vienna *X-ray spectrometry* **39/2** pp 98-102

IOP Conf. Series: Materials Science and Engineering **37** (2012) 012009 doi:10.1088/1757-899X/37/1/012009

Use of x-ray fluorescence and diffraction techniques in studying ancient ceramics of Sri Lanka

B S B Karunaratne
Postgraduate Institute of Science,
University of Peradeniya, Peradeniya,
Sri Lanka.

E-mail: director@pgis.lk and bsbk@pdn.ac.lk .

Abstract. Ceramics were produced for centuries in Sri Lanka for various purposes. Ancient ceramic articles such as pottery, bricks, tiles, sewer pipes, etc, were made from naturally occurring raw materials. Use of X-ray fluorescence (XRF), X-ray diffraction (XRD) and Scanning Electron Microscopy (SEM) in characterizing of two ancient ceramic samples from two different archaeological sites in Sri Lanka is presented. The information obtained in this manner is used to figure out the ancient ceramic technology, particularly to learn about the raw materials used, the source of raw materials, processing parameters such as firing temperature or binders used in ceramic production. This information then can be used to explore the archaeometric background such as the nature and extent of cultural and technological interaction between different periods of history in Sri Lanka.

1. Introduction

Ceramics have been produced from naturally occurring raw materials for centuries in Sri Lanka for various purposes. Although study in ancient ceramics has been an important area in archeology in Sri Lanka for many years [1- 4], a very little research has been done by natural scientists in this regard. During the last decades, X-ray techniques have gained growing interest globally among the natural scientists as a nondestructive analytical tool of studying archeological objects [5]. The ancient civilization knew that clay materials become plastic when water was added and could be moulded into different shapes. These green shapes were then dried in the sun and fired at high temperature mainly using fire wood. Many raw materials used by our ancestors are still utilized by us today in the traditional ceramic industry.

The clay minerals are mainly hydrated alumino-silicates. During the firing process polymorphic transformation of some raw materials (e.g. quartz), decomposition of crystalline compounds to form new phases and a variety of other microstructural changes can occur [6]. In most silicate systems formation of a silicate liquid at firing temperature is an important event in relation to densification. Upon cooling the ceramic body, this liquid will remain as a glassy phase in the microstructure. The compositions of the glassy phase and the crystalline phase depend on the initial overall composition of the raw materials, fineness and the firing temperature. Microstructural information together with the analytical data can thus give some idea of the manufacturing process such as the firing temperature, firing atmosphere and the initial composition of the ancient ceramic items.

2. Experimental

A preliminary study was carried out by us using X-ray fluorescence (XRF; Fischerscope DF500FG-456,) and X-ray diffraction (XRD; Siemens D50000) techniques in conjunction with Scanning Electron Microscopy (SEM; JEOL JSM 6320F) in order to characterize two ancient pottery samples from two different archaeological sites in Sri Lanka (Sample 1-Anuradapura, "Citadel" 4th or 3rd century BC, sample 2- Pinwewa-Galsohonkanatha - Megalithic burial site-approximately 4th century BC). The appearance of the interior and the exterior surfaces were somewhat different (figure 1). In both samples the external surfaces appeared black and the internal surfaces appeared brown. The surfaces of the samples were cleaned by grinding off the surfaces using SiC papers before the investigations.

Figure 1. Photographs of pottery samples
(a) Interior of sample 1 (brown)
(b) Exterior of sample 1 (black)
(c) Cross section of sample 1
(d) Interior of sample 2 (brown)
(e) Exterior of sample 2 (black)

The optical microscopy showed that the quartz crystals of varying size from 0.1 mm to 1 mm were dispersed in the ceramic samples. It was also observed that, in the second sample the fraction of coarse particles of quartz crystals were higher. In general the appearance of sample 2 was much more rugged than that of sample 1. Scanning Electron Microscopy (SEM) was used to compare the grain size of the two samples. The elemental analysis was carried out qualitatively by X-ray fluorescence technique and the phase distribution was determined by X-ray diffraction (XRD).

3. Results and discussion

The XRF analysis showed that samples 1 and 2 have a similar elemental distribution (figure 2 and figure 3). For comparison a present day pottery sample was also analyzed (figure 4). The analysis showed that the elemental composition of the samples is similar. Usually the elements are present in oxide form in raw materials and they react chemically and form different chemical compositions during firing [6]. As mentioned earlier the appearance of the internal and external surfaces of the archaeological samples was different and XRD analysis was carried out separately on both sides. The surfaces of the samples were smoothly grounded and made flat before the investigations. It is also worth noting that the XRD patterns of the interior and the exterior surfaces of the samples were somewhat different (figure 5 and figure 6). This may be due to deliberate addition of different raw materials (rutile rich ilmanite containing clay) as a coating or a glazing to the exterior side of the ceramic pots.

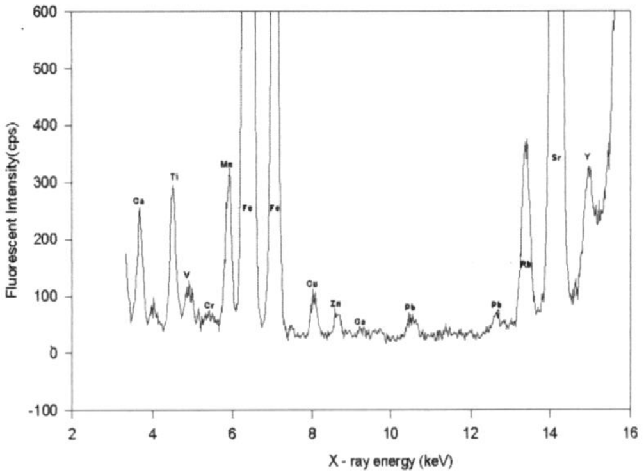

Figure 2. The elemental composition of the sample 1 (XRF spectrum)

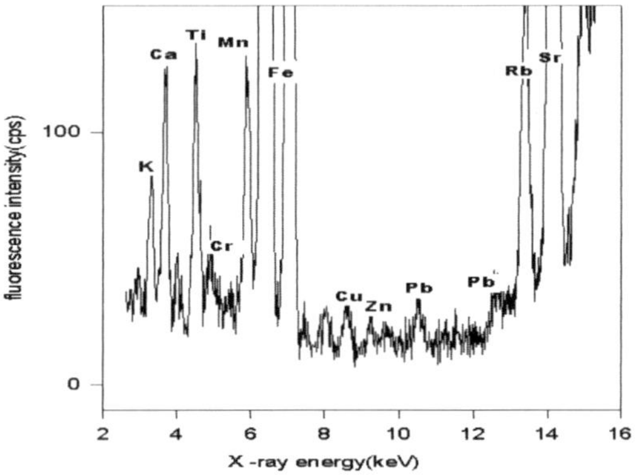

Figure 3. The elemental composition of the sample 2 (XRF spectrum)

XRD patterns of both samples 1 and 2 showed the presence of Quartz in the pottery samples. Quartz had obviously come from the raw materials. However, the XRD pattern (figure 7) of sample 2 only indicated presence of mullite in the sample. Conversion of clay compounds to form mullite can occur if the firing temperature was above 800 ^{o} C [7]. As such, the presence of mullite peaks in the XRD pattern of sample 2 indicates that the firing temperature had exceeded 800 ^{o}C. However, in the case of sample 1, mullite peaks were absent in the XRD pattern (figure 6) and hence the firing temperature could be below 800 ^{o}C.

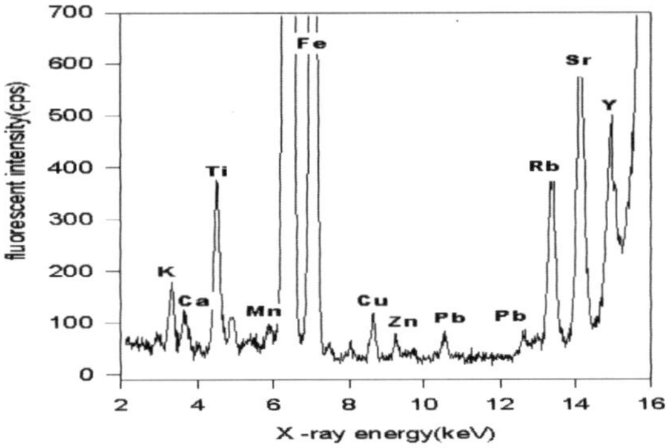

Figure 4. The elemental composition of a present day pottery sample

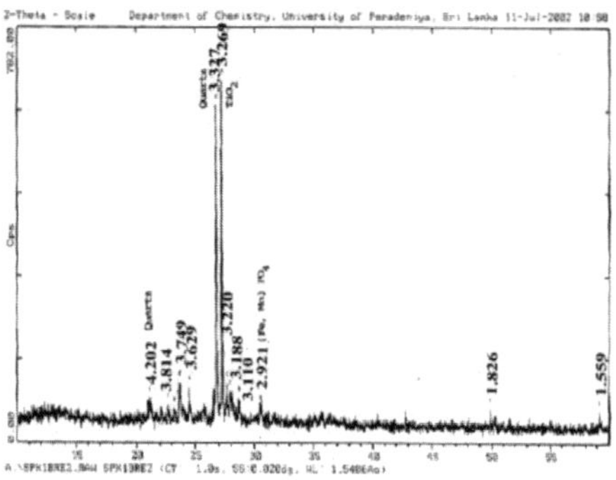

Figure 5. The phase distribution of exterior side (black colour) of the sample 1

SEM studies showed that the grains in sample 2 were much coarser than the grains in sample 1 indicating an extended firing duration of sample 2. It is expected to extend the study to obtain more data and to study more samples. A large quantity of data obtained in this manner would be useful in understanding the ancient ceramic technology, particularly the raw materials used, the source of raw materials, processing parameters such as firing temperature, firing duration and binders used in ceramic products. This information can then be used to explore the archaeometric background such as the nature and extent of cultural and technological interaction between different periods of history of Sri Lanka. This preliminary study has shown that ancient pottery in Sri Lanka was made using similar raw materials and firing temperatures around 800 °C. In conventional pottery industry, the firing is done using fire wood and the temperature in the kiln is around 800 °C.

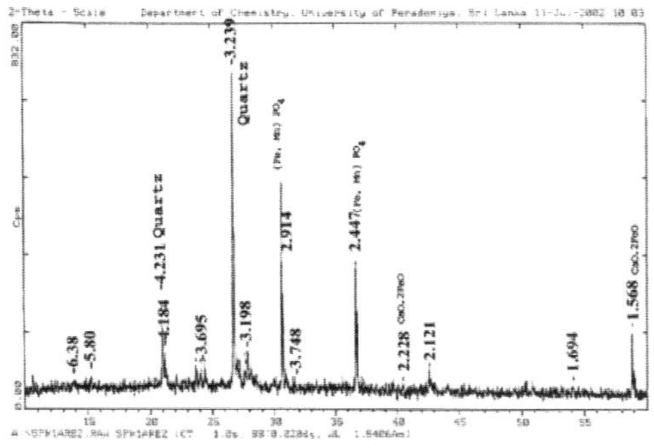

Figure 6. The phase distribution of interior side (brown colour) of sample 1

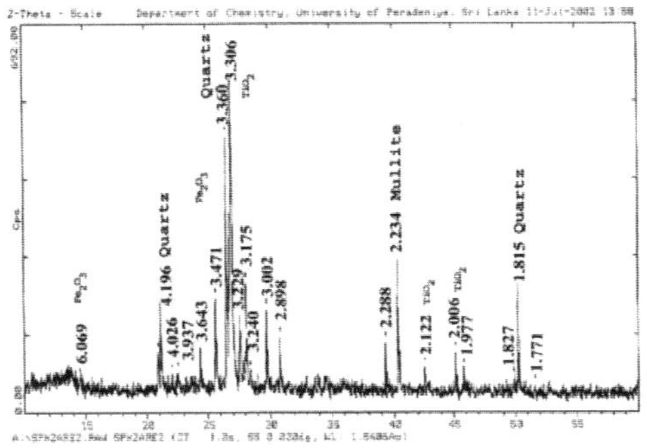

Figure 7. The phase distribution of the interior side brown colour) of sample 2

The scanning micrographs revealed that the grains in sample 2 were coarser than the grains in sample 1. The samples were porous and there were frequent cracks visible in the microstructure. A convenient method of gathering microstructural (phase distribution) and chemical information of a sample is the use of an SEM equipped with a backscattering detector and a micro-analyzer. However, for the present work this technique was not used due to the unavailability of such a facility. It is planned to do some extensive study and analyze sample elemental distribution quantitatively using more samples obtained from different parts of Sri Lanka. This data are useful in identification of the source of raw materials and hence the location of raw materials obtained for manufacturing ancient pottery.

4. Conclusions

This preliminary study has shown that ancient pottery (fourth century BC) in Sri Lanka, was made using similar raw materials and similar firing techniques used today in the conventional pottery industry in Sri Lanka.

Acknowledgements

I wish to thank Prof. Surdharshan Seneviratne of the Department of Archaeology, University of Peradeniya for supplying the pottery samples for this study.

References

[1] Coningham R A E, Gunawardhana P, Adikari G, Katugampola M, Simpson I and Young R 2006 *South Asian Studies* **22** 53
[2] Ratnayake H and Perera R M M B 1984 Classification of pottery in Jetavana vihara project Anuradapura *First archaeological excavation and research report* 59
[3] De Silva N and Dissanayake R B 2008 *A catalogue of ancient pottery from Sri Lanka* Colombo Postgraduate institute of archaeology Sri Lanka
[4] Coningham R A E 1999 *Anuradhapura Volume 1 Society for South Asian Studies (British Academy) Monograph no. 3* Archaeopress Oxford
[5] Schreiner M, Frühmann B, Jembrih-Simbürger D and Linke R 2004 *Adv. in X-ray Analysis* **47** 1
[6] Lee S, Kim Y J and Moon H S 1999 *J. American Ceram. Soc.* **82** 2841
[7] Blanchart P, Deniel S and Tessier-Doyen N 2010 *Advances in Science and Technology* **68** 31

IOP Conf. Series: Materials Science and Engineering **37** (2012) 012010 doi:10.1088/1757-899X/37/1/012010

Pots, plates and provenance: sourcing Indian coarse wares from Mleiha using X-ray fluorescence (XRF) spectrometry analysis

A Reddy[1], AG Attaelmanan[2] and M Mouton[3]

[1] Historic Environment Department, Abu Dhabi Tourism and Culture Authority, P.O. Box 2380, Abu Dhabi, United Arab Emirates
anjana.reddy@adach.ae

[2] Applied Physics Department, University of Sharjah, United Arab Emirates
aattaelmanan@sharjah.ac.ae

[3] CNRS / UMR 7041 Maison de l'Archéologie et Ethnologie, 21 Allée de l'Université, 92023 Nanterre cedex, France
michel.mouton@mae.cnrs.fr

Abstract.
The identification of more than 25% of the pottery sherds from the late PIR.D period (ca. 2nd - mid. 3rd c. AD) assemblage from the recently excavated building H at Mleiha as Indian is based on form and fabric, but using only visual assessment. Petrographic analysis of the fabrics can provide more precise indicators of the geographical origin of the wares. In this study, a total of 21 sherds from various key sites in Western India were compared with 7 different 'Indian' coarse-ware vessels sampled at Mleiha using X-ray fluorescence (XRF) spectrometry. The analyses were conducted on powdered samples collected from the core of each sherd. Each sample was irradiated for 1000 seconds using a 1.2 mm diameter X-ray beam. The resulting spectra were used for quantification of the X-ray intensity and elemental concentration. Levels of correlation in the elemental ratios of the sherds were statistically tested using an F-test as well as a Chi-test. Initial review of the XRF results indicates that the Maharashtra and Gujarat regions of India are probable source areas for at least two of the types of wares. Collection of additional samples from these areas and other regions of India, and further statistical analysis through methods such as Principal Component Analysis will help to isolate groups of wares from India and correlate them with types of vessels imported into the Oman peninsula in antiquity.

1. Introduction: Indian pottery as an indicator of overseas trade at Mleiha.

In recent research at Mleiha (Emirate of Sharjah, United Arab Emirates), fieldwork has focused on the later period of occupation of the site, period PIR.D (mid. 2nd - mid. 3rd c. AD). Mleiha was involved in long-distance trade from the early phases of its occupation (late 3rd c. BC) but in the later period the Indian pottery types increased in number, providing evidence of systematic trade across the Indian Ocean [1], [5]. These goods were part of a more general lucrative trade between the Indian Ocean regions and the Roman world at least from the 1st c. BC to the 3rd c. AD, including for instance, as far as pottery is concerned, transport vessels from South Arabia, glazed table wares from Mesopotamia, and cooking wares and storage jars from India [7].

On the base of the macroscopic observations, pottery from the Indian subcontinent represents 28.35% of the ceramic assemblage at building H, excavated during the two last seasons of work at Mleiha (2010-2011). The samples analysed and presented in the present paper focus on some very well represented wares showing morphological parallels with western Indian sites, and more precisely from the modern states of Maharashtra and Gujarat. After taking into consideration parallels in vessel shapes, the next step should be chemical characterisation. X-Ray Fluorescence Spectrometry analysis involving both mineralogical and geochemical methods, have been applied to samples from the sites in Western India in order to compare the results with supposed Indian ceramics collected at Mleiha. The primary objective was to establish the eventual correlation with pottery production sites, regions and / or trading / supply centres in India in order to establish the likely geographic origin of these wares.

More importantly from an archaeological point-of-view, the samples were selected from a range of Early Historic sites which emerged in Western India, particularly Gujarat, more than a millennium after the decline of the Harappan harbours [3]. These include the sherd samples selected from the sites of Dwarka, Prabhas Patan and Padri - signifying the continuation of sea-borne connections that evolved in the Bronze Age between Mesopotamia, the Oman peninsula and India. From Maharashtra, the selected sites have a strong historic significance, mentioned in the Periplus as important hinterland trade centres which transported goods (including pottery) via caravan routes (e.g. the sites of Ter - mentioned as 'Tagara' in the Periplus - Nevasa, Junnar and Nasik) and ultimately participating in the Indian Ocean trade involving the Arabian peninsula, South Arabia and the Red Sea sites in the early centuries AD. The present selection of sherd samples from India were collected from levels dating to the Early Historic period in India; From Maharashtra - Ter (Period II - c. 1st century AD - 3rd century AD), Nevasa (Period V - c. 50 BC - 200 AD), Nasik (Period IIB - c. 200 BC - 50 AD) and Junnar (Phase I & II - c. between 2nd century BC and 2nd century AD) and from Gujarat – Dwarka (Period II – c. 1st – 4th century AD), Padri (1st century BC), Prabhas Patan (Period IV - 0 AD - 6th century AD) and Gorasa Hanuman (surface site - Early Historic Period) (Fig. 1).

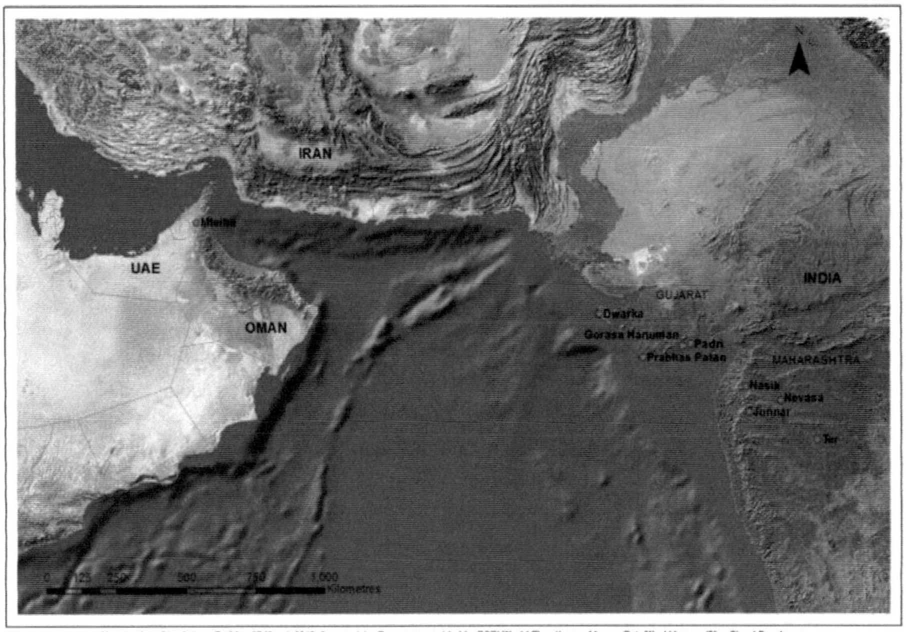

Map produced by Anjana Reddy - 27 March 2012. Source data: Basemap provided by ESRI World Elevation and Image Data/World Images/2km Cloud Free Image.
Collection of pottery samples - location of sites in UAE (Mleiha) and India (within Gujarat and Maharashtra). Anjana Reddy.

Figure 1 – Map showing the location of sites in UAE (Mleiha) and India (within Gujarat and Maharashtra) from where the pottery samples were collected. (Source data: Basemap provided by ESRI World Elevation and Image Data/World Images/2km Cloud Free Image)

1.1. Mleiha Indian pottery: General discussions on vessel form and fabric

The Indian pottery from Mleiha has a form - fabric combination that is easily recognizable as distinct from the Mesopotamian, Egyptian, South Arabian and Arabian Ware in the ceramic assemblage at Mleiha [7]. However, in the absence of Indian fine wares, the analysis is restricted to samples of coarse wares from Mleiha.

Parallels with ceramics from India can be listed in the Mleiha repertoire based on both unusual as well as typical Indian pottery forms. These include *handis* (Indian term for carinated cooking vessels), carinated plates, lamps or lids, flasks or funnel-mouthed vessels, globular vessels with ridged carination and large storage jars or cooking pots. The second, more important determinant in the identification of Indian pottery is fabric. In earlier publications, coarse-ware fabrics of possible Indian origin from the Mleiha ceramic assemblage have been broadly defined as "brown ware with chalky / shelly / sandy grits" [1], [2]. We can now distinguish a brown reddish sandy ware with white chalky inclusions from a brown reddish ware with shelly temper. In addition to this, another fabric of likely Indian source is mentioned as a red ware with a black core and micaceous temper [1].

2. Materials and Methods

2.1 Samples

Sherd samples from India were obtained by both excavation and surface collection from coastal and hinterland sites including Ter, Nevasa, Nasik and Junnar in Maharashtra and Dwarka, Padri, Prabhas Patan and Gorasa Hanuman in Gujarat. The selection was based partly on published pottery illustrations from excavation reports and partly on familiarity with Indian ceramics from these sites (by one of the authors).

A total of 21 sherds from various key sites in Western India were compared with 7 samples collected from phase PIR.D at Mleiha, representing the three main types of wares (as described in section 1.1) with shapes paralleled to industries from Western India (ML-10014 P, ML-10043 Q, ML H-5004 R, ML H-5008 S, ML H-5002 T and ML-10068 U), and a unique sample (ML-2280) of a very characteristic industry from the early PIR.A period at Mleiha (layer 3rd - mid 2nd c. BC) which can be compared with productions from the Early Historic period in Maharashtra.

2.2 XRF analysis

The analysis were conducted using an X-ray Analytical Microscope (XGT 7200, Horiba, Japan), on powder samples collected from the clay matrix core of the sherds, thereby excluding any treatment of surface or temper so that the correlation between the various sherd samples is based on the geo-chemical and elemental concentration. Each sample was analyzed under vacuum for 1000 seconds using a 1.2 mm X-ray beam generated from a Rhodium (Rh) X-ray tube operating at 50 kV and 0.5 mA. The collected spectra were used to identify and quantify elements present in the pottery fabric.

On a routine basis the following 16 elements were detected on each individual sample: Aluminum (Al), Silicon (Si), Sulfur (S), Potassium (K), Calcium (Ca), Titanium (Ti), Chromium (Cr), Manganese (Mn), Iron (Fe), Nickel (Ni), Copper (Cu), Zinc (Zn), Rubidium (Rb), Strontium (Sr), Zirconium (Zr), and Tin (Sn). The mass and molecular percentage of the major and trace elements of each sample were calculated. To begin with, X-ray intensity values were used to test the correlation levels in the sherds' geochemical origin. The samples were analysed and classified by statistical means using the Chi-test.

2.3. Chi-Test

CHI-TEST returns the value from the chi-squared (χ^2) distribution for the statistic and the appropriate degrees of freedom. χ^2 tests can be used to determine whether hypothesized results are verified by an experiment.

The χ^2 test first calculates a χ^2 statistic using the formula:

$$\chi^2 = \sum_{i=1}^{r}\sum_{j=1}^{c}\frac{\left(A_{ij} - E_{ij}\right)^2}{E_{ij}}$$

where:
A_{ij} = actual frequency in the i-th row, j-th column
E_{ij} = expected frequency in the i-th row, j-th column
r = number or rows
c = number of columns
A value of 1 confirms complete correlation between the two tested variables, while a low value of χ^2 is an indicator of independence. As can be seen from the formula, χ^2 is always positive or 0, and is 0 only if $A_{ij} = E_{ij}$ for every i,j.

3. Analytical Results

X-ray fluorescence spectra (Fig. 3) were de-convoluted using the Fundamental Parameter method to quantify elements present in the samples in their oxide form (Al_2O_3, SiO_2, SO_3, K_2O, CaO, TiO_2, Cr_2O_3, MnO, Fe_2O_3, NiO, CuO, ZnO, Rb_2O, SrO, ZrO_2, SnO_2). Concentrations within the different samples varied considerably but some samples seem to have similar elemental compositions as shown in tables 1 and 2.

Statistical results (table 3), indicate that out of the seven Mleiha samples, 2 samples (ML 2280 and ML H5004 R) have the highest correlation values with 11 western Indian samples based on elemental concentrations and geo-chemical composition (Fig.2).

Figure 2 - Photographs of the Mleiha pottery samples ML H5004 R and ML2280 (from left to right) which show highest correlation values with the Indian samples based on elemental concentrations

4. Discussion and conclusion: Significance of the XRF results on the Indian pottery samples from Mleiha and future prospects

Only two sherds from Mleiha have strong correlations of chemical composition with the sherd samples from sites in Maharashtra and Gujarat, indicating more than a 90% probability that they are from the same environment. Out of the 21 samples selected in India, 13 samples showed a high correlation with these two Mleiha samples (ie ML 2280 and ML H5004 R). From Gujarat 4 samples, from a total of 8 samples, (Dwarka CRW 2, Dwarka Red Slipped, Padri Red Slip and Gujarat Prabhas Patan RPW) indicate parallels with the Mleiha samples. With regard to Maharashtra, out of a total of 13 samples selected, 9 samples (Junnar Common CRW, Nevasa CRW NVS, Ter SVG-34, Nevasa NVS Sa 341 RCW, Nevasa CRW, Junnar Red Slipped, Nasik Red Slipped, Nevasa Black Ware and Nevasa RPW I D2) correlate with the 2 samples from Mleiha (Fig. 3).

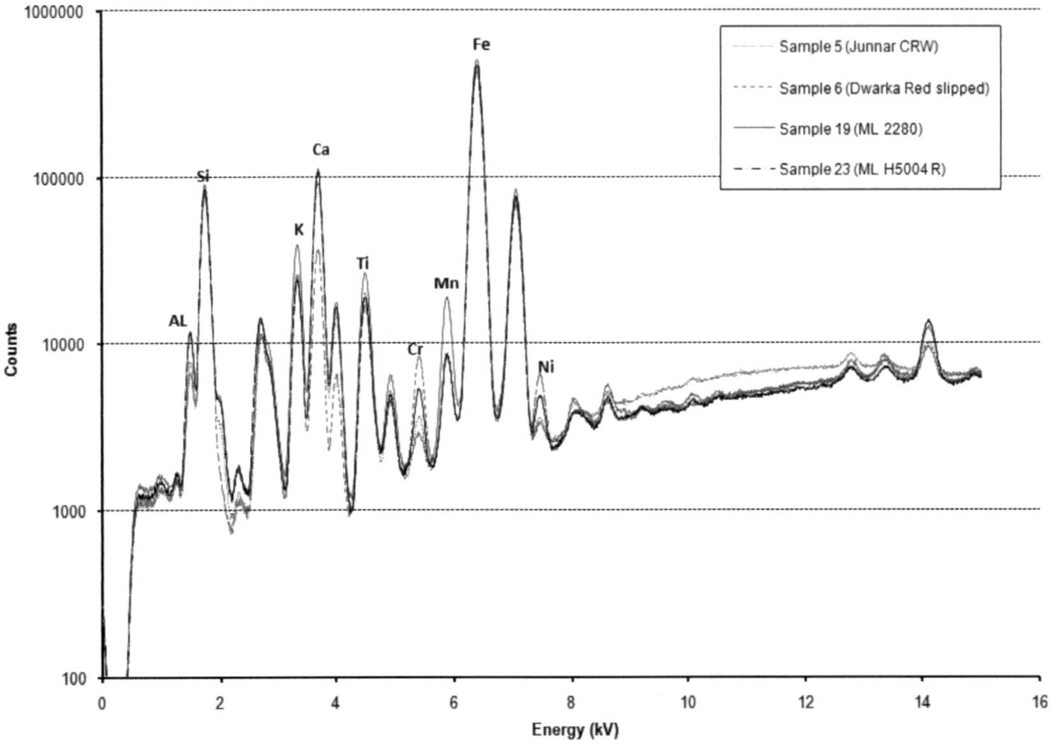

Figure 3 – Examples of XRF Spectra plots of photon counts (log) versus photon energy (keV) that show how well Indian pottery samples (5 & 6) correlate with Mleiha pottery samples (19 & 23).

Table 1. Percentage Elemental compositions of Mleiha sample ML H5004 R and Western Indian samples. Other trace elements also present in but not shown here Rb_2O, SrO, ZrO_2, SnO_2.

Sample	Al_2O_3	SiO_2	SO_3	K_2O	CaO	TiO_2	Cr_2O_3	MnO	Fe_2O_3	NiO	CuO	ZnO
ML H5004 R	12.45	62.01	0.29	2.6	9.98	1.37	0.14	0.18	10.77	0.05	0.01	0.02
Junnar common CRW	9.43	63.08	0.19	3.01	9.83	1.61	0.08	0.21	12.44	0.02	0.02	0.02
Dwarka Red Slipped	12.68	69.21	0.17	3.23	3.61	1.14	0.24	0.16	9.38	0.08	0.01	0.01
Nevasa CRW NVS	10.03	58.11	0.32	3.42	10.35	2.03	0.09	0.2	15.26	0.03	0.02	0.03
Padri Red Slip	11.2	59.08	0.1	1.94	8.18	2.53	0.22	0.27	16.27	0.07	0.01	0.01
Dwarka CRW 2	11.6	61.28	0.52	3.84	10.24	1.07	0.06	0.22	11.04	0.02	0.01	0.02
Nevasa NVS Sa 341 RCW	9.85	57.75	0.41	3.77	12.53	2.17	0.09	0.28	13.01	0.02	0.02	0.02
Gujarat prabhas patan RPW	14.61	63.75	0.03	3.75	6.51	1.08	0.07	0.09	10.03	0.02	0.01	0.01
Nevasa CRW	9.73	56.45	0.32	2.63	11.64	2.22	0.18	0.35	16.25	0.06	0.02	0.03
Nevasa RPW I D2	10.43	66.02	0.29	4.65	6.86	1.21	0.04	0.16	10.22	0.02	0.01	0.01
Nasik Red Slipped	11.04	59.46	0.11	2.64	6.39	2.47	0.09	0.28	17.33	0.03	0.02	0.02
Nevasa Black Ware	11.11	58.55	0.18	2.27	7.3	2.53	0.12	0.25	17.49	0.03	0.03	0.02

These results suggest a possible trade relationship between the Oman peninsula and the western Indian regions in antiquity. More precise analysis may isolate the sources of each type of imported vessels. Principal Component statistical analysis is required as well as a combination of the current geo-chemical / XRF investigation methods with the petrographic analysis of pottery samples (X-Ray Diffraction, Thin-Section analysis etc.). The limitations regarding the places of production may be also addressed by focusing on surface slips or pigments rather than clay matrices [6]. That work has to be done on a larger number of samples to determine characteristic groups of each ware

Table 2. Percentage Elemental compositions of Mleiha sample ML 2280 and Western Indian samples. Other trace elements also present in but not shown here Rb_2O, SrO, ZrO_2, SnO_2.

Sample	Al_2O_3	SiO_2	SO_3	K_2O	CaO	TiO_2	Cr_2O_3	MnO	Fe_2O_3	NiO	CuO	ZnO
ML 2280	**7.36**	**60.36**	**0.23**	**4.53**	**11.36**	**2.15**	**0.04**	**0.51**	**13.26**	**0.02**	**0.03**	**0.03**
Junnar common CRW	9.43	63.08	0.19	3.01	9.83	1.61	0.08	0.21	12.44	0.02	0.02	0.02
Nevasa CRW NVS	10.03	58.11	0.32	3.42	10.35	2.03	0.09	0.2	15.26	0.03	0.02	0.03
Ter SVG-34	6.26	58.3	0.2	2.34	11.85	2.53	0.07	0.35	17.88	0.03	0.03	0.04
Junnar Red Slipped	6.88	53.88	0.22	3.28	13.36	2.17	0.15	0.33	19.48	0.04	0.03	0.04
Dwarka CRW 2	11.6	61.28	0.52	3.84	10.24	1.07	0.06	0.22	11.04	0.02	0.01	0.02
Nevasa NVS Sa 341 RCW	9.85	57.75	0.41	3.77	12.53	2.17	0.09	0.28	13.01	0.02	0.02	0.02
Nevasa CRW	9.73	56.45	0.32	2.63	11.64	2.22	0.18	0.35	16.25	0.06	0.02	0.03
Nasik Red Slipped	11.04	59.46	0.11	2.64	6.39	2.47	0.09	0.28	17.33	0.03	0.02	0.02
Nevasa Black Ware	11.11	58.55	0.18	2.27	7.3	2.53	0.12	0.25	17.49	0.03	0.03	0.02
Padri Red Slip	11.2	59.08	0.1	1.94	8.18	2.53	0.22	0.27	16.27	0.07	0.01	0.01
Nevasa RPW I D2	10.43	66.02	0.29	4.65	6.86	1.21	0.04	0.16	10.22	0.02	0.01	0.01

Anal. A.G. Attaelmanan. Element data in weight-percent

Analysis of the samples from Mleiha have shown three groups of related wares that are most probably representative of regional industries (table 4): the two samples discussed above, ML H-5004 R and ML-2280, of brown ware with a gritty temper and reddish slip, are strongly related (0.99); two samples of brown reddish ware with sandy and micaceous temper, ML H-5002 and ML-10014 are also related (1.00) forming a second group; the third group of related wares (0.92, 0.95, 0.96), is represented by the two samples of brown reddish ware with white chalky inclusions, ML-10043 and ML H-5008, together with sample ML-10068 of brown reddish ware with shelly temper. These three groups are clearly different from one another (0.00 to 0.42). They are representative of three different regional industries, from which only one (ML H-5004 and ML-2280) was represented in the collection of samples from India that are at our disposal. The two other groups may originate from other regions of India, as the shapes of the vessels represented in these three groups of wares can be compared and are culturally related.

At this stage of analysis and research we cannot determine the exact provenance of the industries represented at Mleiha. The correlations recorded at present between the Mleiha sherds and samples from Western India do not necessarily indicate the actual pottery production/manufacturing areas within India. Future research based on further sampling from Indian sites will therefore help define or locate the Indian industries more precisely. This requires careful selection of samples (20 to 30 different samples from each industry). The results from the correlations between each group/industry will determine the exact parallels between the Indian wares found in the Oman peninsula with the regional industries within India.

At present, we can hypothesize that the Indian material was imported to Mleiha most probably through Dibba harbour [4] in the north-east of the Oman peninsula. The abundance of Indian domestic vessels in these sites in antiquity leads us to enquire into the likely presence of an Indian population on the coasts of Arabia, at Dibba, Khor Rori (Dhofar), Qani (Hadramawt), and even at inland sites like Mleiha, on the western foothills of the Oman Mountains.

Table 3. Correlation values between the Mleiha and Western Indian samples (1.0 = 100%)

Sample	Correlation Value	Correlating Indian Sample
ML 2280	1.0	Junnar Common (CRW)
	1.0	Nevasa CRW NVS
	1.0	Ter SVG-34
	1.0	Dwarka CRW 2
	1.0	Nevasa NVS Sa 341 RCW
	1.0	Nevasa CRW
	0.99	Junnar Red Slipped
	0.97	Nasik Red Slipped
	0.97	Nevasa Black Ware
	0.97	Padri Red Slip
	0.96	Nevasa RPW I D2
ML H-5004 R	1.0	Junnar common
	1.0	Nevasa CRW NVS
	1.0	Dwarka CRW 2
	1.0	Nevasa NVS Sa 341 RCW
	1.0	Gujarat Prabhas Patan RPW
	1.0	Nevasa CRW
	1.0	Nevasa RPW I D2
	0.99	Dwarka Red Slipped
	0.99	Padri Red Slip
	0.97	Nasik Red Slipped
	0.97	Nevasa Black Ware

Table 4. Correlation between the samples collected at Mleiha building H.

Samples	ML H 5004	ML-2280	ML-10014	ML H-5002	ML-10043	ML H-5008
ML H-5004						
ML-2280	**0.99**					
ML-10014	0.40	0				
ML H-5002	0	0	**1**			
ML-10043	0	0	0	0		

MLH-5008	0.42	0.28	0	0	**0.95**	
ML-10068	0	0.01	0	0	**0.96**	**0.92**

Acknowledgements
The Indian sherd samples are from the collections held by the Department of Archaeology, Deccan College Post-Graduate and Research, Pune - India. Special thanks go to Prof. Vasant Shinde, Dr. Prabodh Shirvalkar and Nilesh Jadhav of Deccan College (Pune), and Dr. Amol Kulkarni of Maharashtra State Archaeology (Mumbai) for providing invaluable information on the Indian pottery from Western India. Comparative samples of pottery from Gorasa Hanuman (Gujarat) were provided by Dr. Prabodh Shirvalkar. We would like to thank also Dr. Abhijit Dandekar and Sachin Joshi of Deccan College, Pune for providing samples and information relating to petrographic analysis of Indian pottery. Special thanks go to Dr. Mark Beech of the Abu Dhabi Tourism and Culture Authority for reviewing the manuscript and providing useful suggestions and comments. Finally we would like to thank Mr. Will Higgs for copy-editing the final draft. XRF analysis were carried out at the National X-ray Fluorescence Laboratory (NXFL), University of Sharjah.

References

[1] Benoist A, Mouton M and Schiettecatte J 2003 *PSAS* **33** pp 59-76.

[2] Cuny J and Mouton M 2009 La transition vers la période sassanide dans la péninsule d'Oman : chronologie et modes de peuplement. In *L'Arabie à la veille de l'Islam* ed J Schiettecatte and CJ Robin (Paris De Boccard) pp 91-133.

[3] Gupta S 1997 *Roman Egypt to Peninsular India: Archaeological Patterns of Trade (1st Century BC-3rd Century AD) PhD thesis (unpublished)* (University of Pune).

[4] Jasim S 2006 *AAE* **17/2** pp 214-237

[5] Mouton M, Tengberg M, Le Maguer S, Reddy A, Soulié D, Bernard V, Le Grand M and Goy J Building H at Mleiha: New evidences on the late PIR.D phase in the Oman Peninsula (2nd-mid. 3rd c. AD) in press *PSAS* **42.**

[6] Peacock E 2009 *Establishing an Elemental Baseline for Sourcing Shell and Shell-Tempered Artifacts in the Eastern Woodlands of North America using Laser Ablation-Inductively Coupled Plasma-Mass Spectrometry (LA-ICP-MS):* Final Report submitted to the National Park Service, National Center for Preservation Technology and Training for Award # MT-2210-07-NC-08.

[7] Tomber R, Cartwright C and Gupta S 2011 *J. Arch. Science* **38** pp 360-366.

IOP Conf. Series: Materials Science and Engineering **37** (2012) 012011 doi:10.1088/1757-899X/37/1/012011

APPLICATION OF X-RAY AND NEUTRON TOMOGRAPHY TO STUDY ANTIQUE GREEK BRONZE COINS WITH A HIGH LEAD CONTENT

M Griesser[1], R Traum[2], K Vondrovec[2], P Vontobel[3], E H Lehmann[3]

[1] Conservation Science Department, Kunsthistorisches Museum, Burgring 5, A-1010 Vienna, Austria, email: martina.griesser@khm.at
[2] Coin Cabinet, Kunsthistorisches Museum, Burgring 5, A-1010 Vienna, Austria, email: rene.traum@khm.at, klaus.vondrovec@khm.at
[3] Paul Scherrer Institute (PSI), Spallation Neutron Source Division, CH-5232 Villigen PSI, Switzerland, email: peter.vontobel@psi.ch, eberhard.lehmann@psi.ch

Abstract
Highly leaded bronze coins of the Coin Cabinet of the Kunsthistorisches Museum (KHM) show progressive corrosion as a result of unfavourable storage conditions within historic wooden cases. In connection to a research project concerning the preservation and conservation of the antique coins the causes for the sometimes severe corrosion were studied by different analytical techniques. Radiography and tomography investigations using neutrons and X-rays were performed at the Paul Scherrer Institute, i.e. the enrichment of lead in the interior of the objects was studied in a non-destructive manner. The tomography results obtained show that in addition to the lead rich areas on the obverse and reverse of the coins (often already clearly visible on the surface due to the formation of white corrosion products) a varying number of lead containing inclusions could be detected within the antique bronze coins. In addition, some information on their casting technique could be gained.

Introduction

The Coin Cabinet of the Kunsthistorisches Museum (KHM), Vienna, is one of the largest in the world. It also holds a large collection of ancient Greek bronze coins which were minted during the Roman imperial time. These so-called Greek imperials were produced in the eastern, Greek speaking provinces of the Roman Empire up to c. 280 AD. Other than in the Roman Imperial coinage blanks made of highly leaded bronze alloy were used frequently. Those coins were issued by a great number of municipalities and were primarily used as local currencies; therefore, also their images are of more local character and provide us with unique historic and otherwise unattested information of their cities.
Only a tiny fraction is on display in the permanent exhibition at the KHM, the major part of the collection is stored in wooden storage cabinets (Fig. 1). Due to the release of organic acid pollutions [1] from the wood and together with other air pollutants (i.e. SO_2, NO_2, NO_x, O_3) severe corrosion develops on the antique bronze coins showing a high lead and/or tin content [2-4]. A white powder of at first lead acetate and later on lead carbonate develops during this process. The coins start to show some points of whitish, powdery corrosion products on the surface (Fig. 2). For single objects the corrosion even leads to a complete destruction of the coin's core, by only leaving an intact outer shell of metal and/or patina. When this shell breaks in certain areas it reveals a completely reacted inner part containing only the corroded powdery material [5] (Fig. 2).

Analysis and results

Within a four years research project the corrosion phenomena of 1,200 coins were studied by different non-destructive analytical techniques to better understand the development of corrosion and to enhance the preservation of the antique coins.

Fig. 1 Historic wooden storage cabinets still in use in the Coin Cabinet of the KHM.

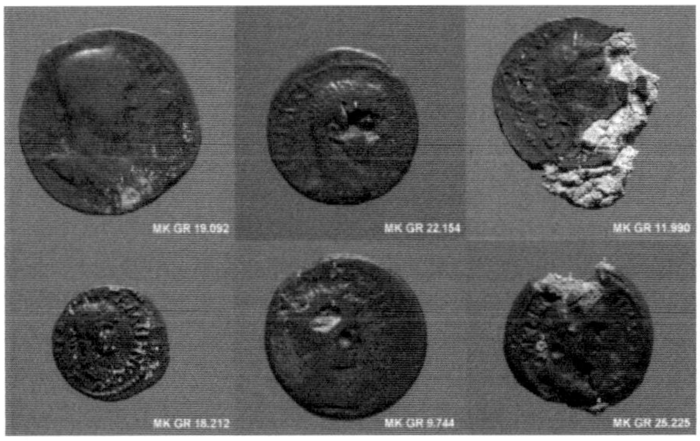

Fig. 2 Different states of typical corrosion phenomena visible on Greek imperials.

UV-fluorescence microscopy was performed in the restoration workshop of the Coin Cabinet using a SteREO Lumar.V12 Zeiss fluorescence microscope equipped with four different UV to red light wavelength filters. Due to their high fluorescence in UV light the formation of mainly carbonates could be observed in the corroded areas at the surfaces of the coins investigated (Fig. 3).

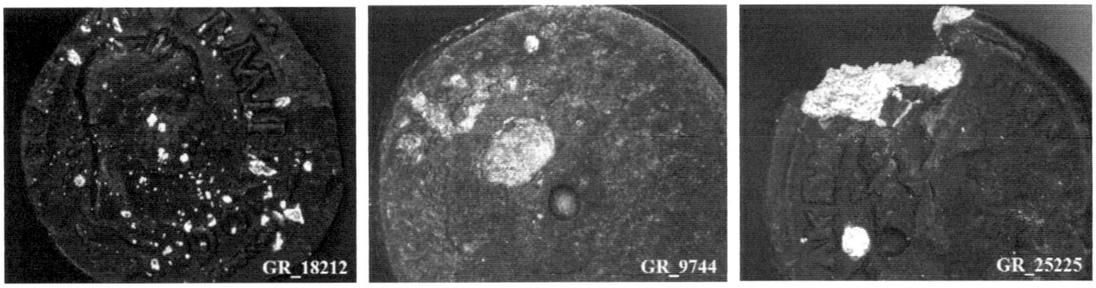

Fig. 3 UV fluorescence microscopy shows the formation of highly fluorescing carbonates.

Scanning Electron Microscopy with Energy Dispersive X-ray detection (SEM-EDX; Philips XL 30 ESEM; 20 kV) proved the presence of mainly lead containing corrosion products (Fig. 4). Whereas the base alloy contains mainly copper (Cu) and lead (Pb) as well as a low amount of tin (Sn), the white corrosion product is clearly Pb based.

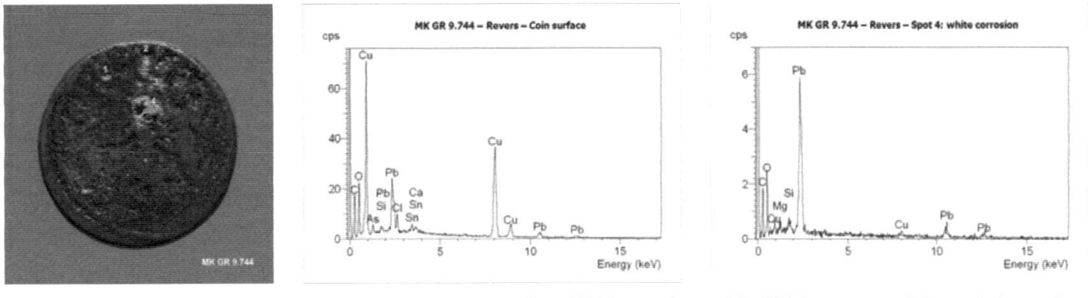

Fig. 4 Example for the SEM results: Revers of coin GR_9744 together with SEM spectra of the coin's surface (bronze) and the white corrosion spot no. 4 in the centre of the upper half of the coin (Pb based corrosion).

To further investigate the phases present in the corroded parts of the coins neutron diffraction studies were performed at the GEM and POLARIS diffractometers at the Rutherford Appleton Laboratories, spallation source ISIS, Oxfordshire, Great Britain, for 20 selected objects with different states of corrosion. These investigations confirmed a high lead content of between 20 and over 30 % Pb and a typical tin content of up to 8 % Sn in the bronze alloys. Unfortunately the evaluation of the corrosion phases is still in progress.
In addition, for the same group of objects X-ray (250 kV) and thermal neutron tomography investigations were performed at the NEUTRA beam line, neutron spallation source SINQ, of the Paul Scherrer Institute (PSI) using a max. image size of 5 x 5 cm^2 and a resolution of 50 μm (for experimental details see [6]). To get a first overview of the inhomogeneity of the coins X-ray and neutron radiographies were taken from all objects (Fig. 5). Very different results could be obtained from the radiographies for the single objects. For some of them nearly the same image could be detected using either X-rays or neutrons, for others the images looked quite different, e.g. much more information on lead (Pb) containing inclusions within the coins could be seen as black areas in the X-ray image than in applying neutron radiography (Fig. 5, left and middle). Sometimes also the opposite result was detected, i.e. more information was to be seen in the neutron image than using X-ray radiography. These differing results depend on Pb being strongly attenuating X-rays due to the many electrons (Z=82) in its atomic shell. Coins with a high Pb content and/or comparatively thick coins are, therefore, less transparent for X-rays than those with a lower Pb content or thinner coins. Pb aggregates (inclusions) within objects show clearly in X-ray, if their thickness is large compared to the object's thickness or if the homogeneously distributed Pb content in the matrix is low. On the other hand, for neutrons Pb is more transparent than e.g. bronze. So in neutron radiographs Pb inclusions are usually less visible than in the corresponding X-ray radiographs. Due to varying distributions of the Pb and the different thicknesses of the coins, exceptions are possible (and were sometimes detected). However, in neutron tomography slices, Pb inclusions are visible for all coins irrespective of their Pb content, because the bronze material is contributing mainly to neutron attenuation.

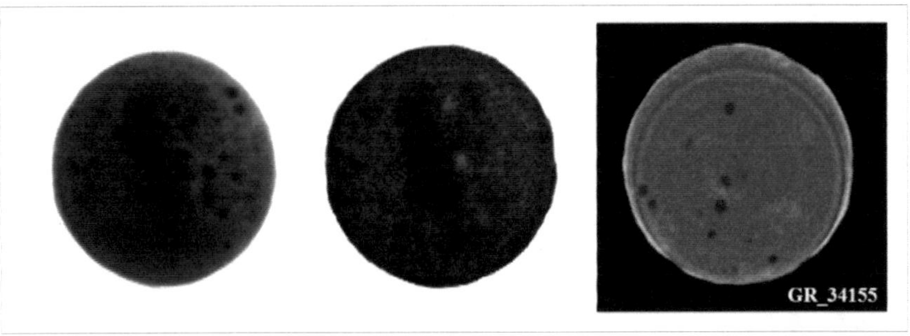

Fig. 5 X-ray radiography (250 kV, left), neutron radiography (middle) and neutron tomography (right) of coin GR_34155 reveal different information concerning its lead rich inclusions.

Following these first investigations the coins were divided in three stacks of six to seven coins, respectively. These stacks were investigated by neutron tomography overnight (360°, 625 images, 60 sec/image), enabling further study of the distribution of the Pb containing inclusions in the core of the objects. For the majority of the coins the presence of two phases within the alloy, one rich in copper the other one rich in lead (darker in the neutron tomography images) could be detected by neutron tomography (Fig. 5, right). As shown on

IOP Conf. Series: Materials Science and Engineering **37** (2012) 012011 doi:10.1088/1757-899X/37/1/012011

selected tomography slices in Fig. 6 the Pb containing inclusions of varying size are inhomogeneously distributed over the width and depth of the coins (typical thickness about 3 mm; Fig. 7).

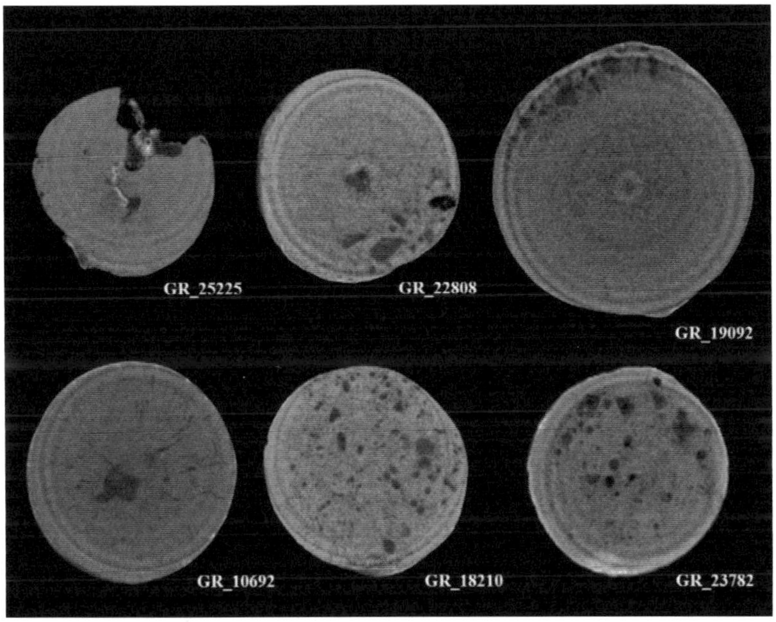

Fig. 6 Casting the flans in vertical moulds leads to a Pb enrichment (darker) at one end of the coins (upper row); for other objects, presumably cast using horizontal moulds, the inclusions seem to be more evenly distributed.

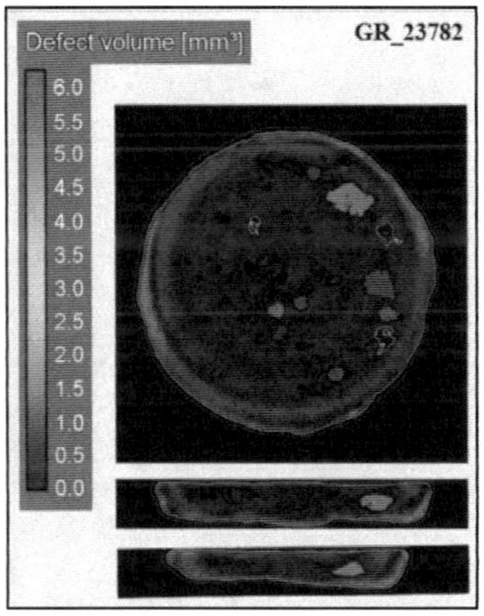

Fig. 7 In this example about 20 % of the coin's volume consists of the Pb rich phase.

During the evaluation of the neutron tomography data it became obvious that additional information on the casting technique – not studied for antique bronze coins so far – can be gained by this non-destructive technique. Due to their high lead content no homogeneous alloy can be formed in the production of the coin blanks used for minting the coins later on. Dependent on the casting technique chosen in antiquity – vertical or horizontal [7-9] – the Pb containing phase can be found in different parts of the object. Using the vertical casting technique a row of upright blanks is received within a closed mould by pouring the hot alloy in from the top. On the other hand, the horizontal casting technique produces lying blanks using open moulds. So the position of the blanks is different between horizontal and vertical casting techniques. Hints for the technique used can sometimes even be seen on the original antique coins, e.g. a nearly perfect round shape for the horizontally cast objects or typical casting marks on two opposite edges of the coins for applying the vertical technique. For the vertically cast blanks a clear enrichment of the Pb rich phase at one end of the coins is often visible (Fig. 6, upper row). Other objects show a more even distribution of the inclusions suggesting a horizontal casting technique. Further studies using self-made replicas applying different ancient casting techniques are still in progress.

The different distribution of the two phases within the coins, together with varying burial conditions and environmental influences, explains the irregular development of corrosion in the core and on the surface of the different coins investigated. To enhance the future preservation of the vulnerable objects a nitrogen flooded metal storage cabinet has been installed in the Coin Cabinet of the KHM recently.

IOP Conf. Series: Materials Science and Engineering **37** (2012) 012011 doi:10.1088/1757-899X/37/1/012011

Conclusion

By applying light and electron microscopy the development of lead containing corrosion products (most probably carbonates) could be confirmed for a set of antique bronze coins with high lead content. Using neutron tomography the development of two different phases during the casting process (one rich in copper and one rich in lead) could be proven and the inhomogeneous distribution of the Pb rich inclusions inside the coins could be shown. The assumption that the scattering of the Pb rich phase is connected to the casting techniques used in antiquity will be further studied using self-made replicas in the future.

References

[1] Hatchfield P B 2002 *Pollutants in the Museum Environment. Practical Strategies for Problem Solving in Design, Exhibition and Storage* (London) pp. 67-74

[2] Grzywacz C M, Tennent N H 1994 Pollution Monitoring in Storage and Display Cabinets: Carbonyl Pollutant Levels in Relation to Artefact Deterioration *Preventive Conservation: Practice, Theory and Research", Preprints of the Contributions to the IIC Ottawa Congress, 12-16 September 1994* (London) pp. 164-170

[3] Tennent N H (1995) Damage to Coin and Medal Collections by Organic Carbonyl Pollutants from Wooden Cabinets *Images for Posterity. The Conservation of Coins and Medals, Papers of the Leiden seminar, 1993* (Leiden) pp. 45-58

[4] Tétrault J, Sirois J, Stamatopoulou E (1998) Studies of Lead Corrosion in Acetic Acid Environments *Studies in Conservation* **43** pp.17-32

[5] Linke R, Schreiner M, Griesser M, Kadlec G, Traum R, Dembski G (2000) Corrosion Processes endanger Collections of Antique Bronze Coins – are they past saving? *International Numismatic Newsletter* **36** pp. 7-9

[6] neutra.web.psi.ch

[7] Pilon F (2003) UN FAIT UNIQUE EN GAULE ROMAINE: LA DECOUVERTE DE MOULES A FLANS MONETAIRES EN PIERRE CALCAIRE *Schweizerische Numismatische Rundschau* **82** pp. 37-60

[8] Elayi J, Elayi A G (1995) Un moule monetaire de l'epoque d'Alexandre Jannee" *Bulletin de la Societe Francaise de Numismatique* **50/1** pp. 1184-1188

[9] Moesta H, Franke P R (1995) Kap.6 Herstellung von Muenzen *Antike Metallurgie und Muenzpraegung. Ein Beitrag zur Technikgeschichte* (Basel) pp. 77-116

Acknowledgements

The authors want to thank W. Kockelmann and R. Smith for performing the investigations at ISIS. This work is based on experiments performed at the Swiss spallation neutron source SINQ, Paul Scherrer Institute, Villigen, Switzerland and has been supported by the European Commission under the 7th Framework Programme through the 'Research Infrastructures' action of the 'Capacities' Programme, Contract No: CP-CSA_INFRA-2008-1.1.1 Number 226507-NMI3 and by the Jubiläumsfonds der Oesterreichischen Nationalbank, Project No. 11990.

IOP Conf. Series: Materials Science and Engineering **37** (2012) 012012 doi:10.1088/1757-899X/37/1/012012

QXRD, XRF and optical microscopy applied to characterization and provenance of ancient ceramics from Haft Teppeh (1500 – 1150 BC), southwest Iran

Mohammadamin Emami[1, 2]

[1] Faculty of Conservation, Art University of Isfahan, P.O.Box 1744, Isfahan, Iran

[2] Institut für Bau- und Werkstoffchemie, Universität Siegen, Paul Bonatz Str. 9-11, 57068 Siegen, Germany

Abstract

Classifications of Archaeological discoveries are an essential feature for better interpretation of the extractive data from a case excavation. The question that will be focused on in this paper is; which of the parameters – raw materials or manufacturing processes – have changed during the time in an ancient archaeological place. The materials used in this research are ceramic pieces from archaeological excavations carried out in Haft-Tappeh, south-western Iran that chronologically belonging to the middle Elamite period (1500 B.C.). Samples classified firstly by their exterior characters. According to obtained results from quantitative X-ray diffraction and phase refining by Rietveld, it has become possible to characterize the data for clustering the samples in different groups. Observation by polarized light microscopy produced additional information about the manufacturing as well as the processing of the different admixtures used by ancient potters. Mineralogical- chemical interpretation of ceramics based on x-ray methods proved a dissimilar thermal behaviour due to different fabrication conditions during the same period. The results provided principally information on the existence of a common raw material reservoir but different manufacturing technique in pottering during middle Elamite period in Haft-Tappeh.

Key Words: Archaeometry, Ceramic, Rietveld Refining Method, Quantitative Powder Diffraction, Polarization Microscopy, Elamite, Haft Tappeh

1. Introduction

In the middle of the 14th century, the southern district of Susiana plain was chosen as the capital of the Elamite Kingdom by the King "Untash Gal" (Ghirshman, 1968, Negahban, 1991). Haft-Tappeh is located about 20 km south east of Susa and is not only important for being the capital of the Elamite dynasty, but also as a technical centre in the antique world due to metallurgical evidences (Oudbashi & Emami, 2010) (Figure 1). Rivers, underground irrigation, hot and dry weather as well as thaw-freezing effects followed by salt recrystallization are among the main effects of corrosion in this region and also cause well known outcrops of clay deposits on the earth (Bullard 1970, Tehrani 1989). Ancient ceramic artefacts from this region are interesting objects to study according to their varieties in shape as well as their construction. For archaeometrically oriented mineralogist, ceramics are manufactured under different pyrotechnological processes (Maggetti & Schwab). Antique ceramics described as the objects with high porosity and clayey matrix with either different or similar kinds of admixtures which sometimes contain a glassy matrix (Salmang & Scholze, 1983). The ceramics considered here proved to have similar exterior shapes and different physical properties depending on their constituent minerals when compared to ancient ware from Mesopotamia (Noll 1990, Emami & Trettin, 2010).

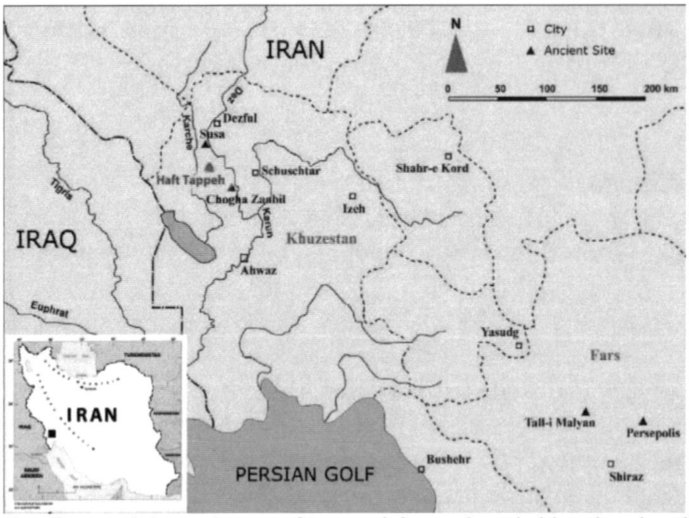

Figure 1. Location of Haft Tappeh in south west of Iran and the roads and cities closed to the sits.

Chemical and crystallographic characterizations of ceramics lead to understanding of the pottering technology used during the specified time period. The ceramic's matrices investigated here are made of minerals that established high temperature firing processes based on their mineralogical constituents. Accordingly, some criterions must be respected for investigation on archaeological ceramics:

(i) Consistency of matrix for interpretation of mechanical properties of ceramics

(ii) Type of binding material such as clayey-, calcareous- and organic based material.

(iii) External shape and form of potteries for interpretation of typology as well as relative chronological identification (Magetti & Gelatti, 1982)

(iv) Characterization of mineralogical constituents for determining the raw material as well as firing temperature (Emami, Volkmar, & Trettin, 2008).

(v) Determination of the temperature due to the firing affects that took place in and on the surface of ceramics (Martineau et al 2007).

2. Material description

50 samples selected for investigation and these are part of the samples dug up in Haft-Tappeh during the excavation in 2002 (Mofidi Nassrabadi, 2003). The ceramics are mostly vase, bowls, jugs and plates. According to their typology the samples described as: (i) Ceramics without any contour on the surface with very simple form. (ii) Ceramics that contained contour as diagonal lines. (iii) Ceramics with simple form and fingerprint or pressure effect on the surface.

3. Experimental

These samples were examined by polarized light microscopy (an instrument from Olympus, XB 51 with AnalySIS Software) and high resolution microscopy methods such as scanning electron microscopy (SEM, from Philips XL 30 with scandium software for better resolution) for mineralogical investigations and characterizations on the crystalline phases. X-Ray Fluorescence (Bruker spectra plus 2008) used for the bulk chemical composition of the major, minor and trace oxides.

IOP Conf. Series: Materials Science and Engineering **37** (2012) 012012 doi:10.1088/1757-899X/37/1/012012

Quantitative x-ray diffraction used for determining the crystalline phases (X-Pert Pro - PANanalytical) and the measured peaks refined with Rietveld method for getting best preferred orientation of the phases. Clustering of analysis carried out for distinguishing the ceramics according to the crystalline phase constituents.

4. Results and Discussion

4.1. Chemical composition of ceramics in system CaO+MgO – SiO_2 – Al_2O_3 and CaO – MgO – (Fe_xO_y+MnO)

The result of archaeological objects analyzed by XRF classified the ceramics in system *CaO+MgO* – *SiO_2* – *Al_2O_3* as calcium-rich, to high calcium-rich ceramics (Figure 2). According to the CSA diagram the results concentrated in quartz – diopside – anorthite line (Noll, 1991). Scattering of the results of chemical analysis depend mainly on the constitution of calcareous material which has been used for pottery making. System CaO+MgO – SiO_2 – Al_2O_3 submit some information about the general chemical composition of ceramics. Scattering of the results along diopside – anorthite line proved firstly that the ceramic's calcareous phases are the main constituents in their texture, and secondly, according to the minimum distribution toward anorthite, the raw materials could acquire from well-known local clay reservoir near Haft-Tappeh. Another useful system for the characterization of the matrix of ceramics is the system CaO – MgO – (Fe_xO_y+MnO) (Fig. 3). According to this diagram the chemical-mineralogical properties of some phases like pyroxene (high temperature phases) could get some more information about the sinter process. The chemical composition of pyroxens varies with respect to the atmospheric conditions in the kiln as well as on raw materials that have been used. The pyroxenes appear throughout the raw material, and compose the most matrices through sintering; secondary pyroxens don't show the gut crystalline aggregation (Figure 3).

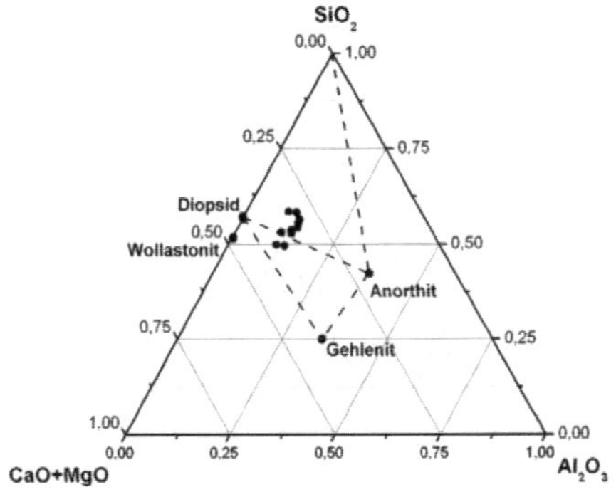

Figure 2. Chemical composition of ceramics with respect to calc constituents in system CaO+MgO-SiO_2-Al_2O_3. Chemical results in W%.

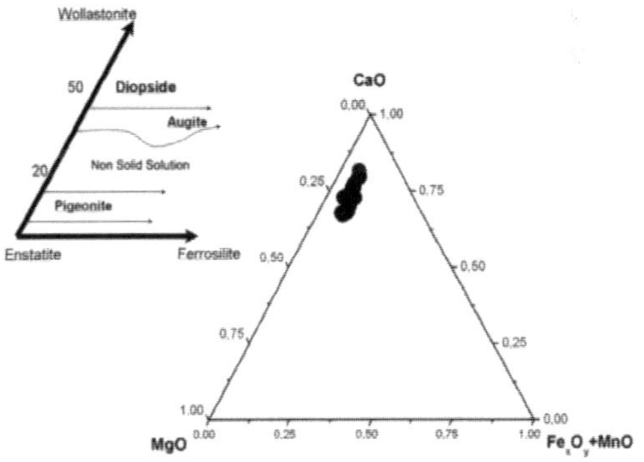

Figure 3. Chemical composition of pyroxene in system CaO – MgO – (Fe_xO_y+MnO). Results in Mol%.

4.2. Phase determination by quantitative X-ray diffraction and Rietveld refining method

Phase determination of the ceramics has carried out by QXRD. The major crystalline phases appear in the ceramic matrix are calcite and quartz (Fig. 4). Quartz reveals coarse grain aggregate and this mineral is mostly used as admixture. Calcite has been used either as flux or appears as secondary crystallized aggregate in the matrix through the firing process. The high amounts of alkalifeldspars imply the usage of potassium-rich raw materials. Gehlenite as well as pyroxene are mainly the reaction's product from clayey raw material. Diopside occurs in Mg-rich matrix that is being rebuilt under high temperature conditions. The high amount of Mg in the texture of ceramics proves that raw materials must derive from dolomitic or a high Mg-calcareous origin. Due to the inhomogenity of the ceramic constituents, all symmetry parameters in the crystal lattice have refined through Rietveld method for getting better crystallographic structure of the calculated phases. Refining of some very fine crystals such as analcime was difficult (trigonal-rhombohedral). This phase detected before the best preferred orientation of quartz (2θ=26°) and also overlapped quartz pick. The qualitative phase investigation in this epoch proves that moscuvite contains 1.4 W% of the ceramic's texture. Muscovite occurs with its best intensity on the surface (025) and comes through the high potassium clayey (illite). In many cases the muscovite is overlapped with the major picks of quartz and diopside. Another important paragenesis in the ceramic occurs through the minerals carrying calcium and magnesium. Calcite, augite, diopside and gehlenite are the main and typical mineral occurrences in the samples from Haft-Tappeh. The occurrences of the phases in a ceramic are dependent on the firing temperature, cooling rates, temperature shrinkage in kiln and fCO_2; while the optical properties of the phases gave the same results. Plagioclase and feldspars are common minerals for estimating the reaction temperature.

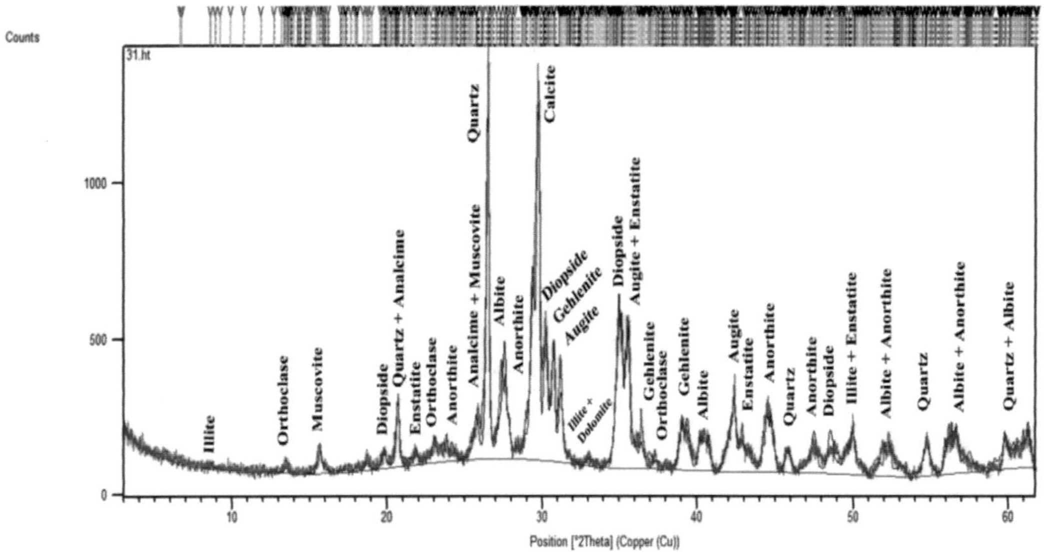

Figure 4. Diffractogram of one of the ceramics analyzed by XRD and refined by Rietveld calculation. Quartz- ICSD 79634, Calcite- ICSD 40544, Anorthite- ICSD 34942, Orthoclase- ICSD 34742, Albite- ICSD 87655, Augite- ICSD 56921, Diopside- ICSD 64977, Enstatite- ICSD 24464, Muscovite- ICSD 202263, Dolomite- ICSD 40971, Illite- ICSD 55333, Gehlenite- ICSD 20392, Analcime- ICSD 87555

4.3. Mineralogical-, petrological observations of ceramics

Matrices in these samples are compact and the porosity is reduced. The porosity in these structures is mainly secondary as they occurred after firing, (since the recrystallization has occurred in the pore or throughout value exchanges). The difference between the textures on these ceramics is therefore due to the quartz – alkalifeldspar – (± plagioclase) and mica. Iron oxides are mainly accumulated in the porosity and cause the commonly observed red color of such objects. The iron rich phase in this figure is an accumulation of colloidal iron-rich components that came from the clay. The new crystallization of iron-rich phases as hematite is a reason for high temperature reaction (Emami & Trettin, 2010). Petrological observations and investigations divided the samples also into two groups with respect to mineralogical constituents in the matrix. The first one mainly consists of aggregates of granite and coarse grain sedimentary stone. The second group consists of calcareous materials. Quartz, in some samples, shows a rounded form with typical undolatory extinction and that could be a reason for transportation and sedimentation of provenance materials (Fiq. 5). The second groups consist of calcite, dolomite and magnesite. Periclase (MgO) appeared through slow oxidation reaction of dolomitic raw materials after decarbonatization. After such reaction calcite decomposed but periclase is accumulated on the surface as a red rim (Fiq. 6). Increasing the Ca-containing minerals will increase the different variations of pyroxene in the matrix; for instance the occurrence of wollastonite, augite and diopside compositions. The stability of such minerals depends on the alkali-oxide/SiO_2 ratio. Plagioclases are the most common minerals in the matrix that form tiny crystals. These aggregates are mostly prismatic with hypidiomorph to xenomorph morphology and also demonstrated alteration effects on the surface as sericitization, that grew along the (010) surface (Fiq. 7). Biotite decomposes through the firing process according to the dehydration reaction by 790 - 840°C. This reaction causes the formation of pertithic structure in biotite and its color will be brighter than natural biotite (Fig. 8).

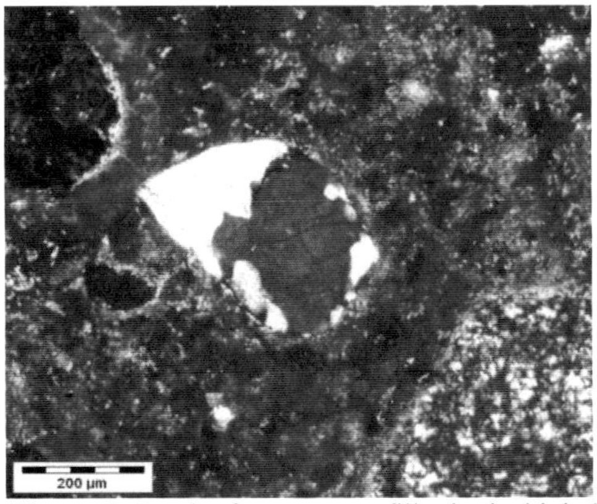

Figure 5. Granite fragments as filler in the high Ca matrix. HT 8013126-x10-2pol

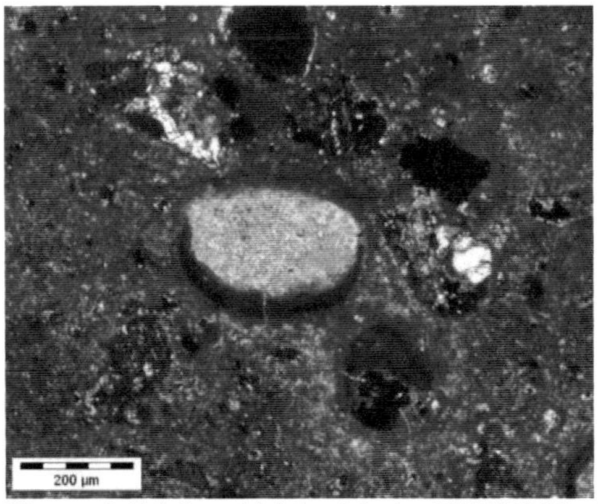

Figure 6. Calcite with Mg-rich rim. HT781256-x5-2pol.

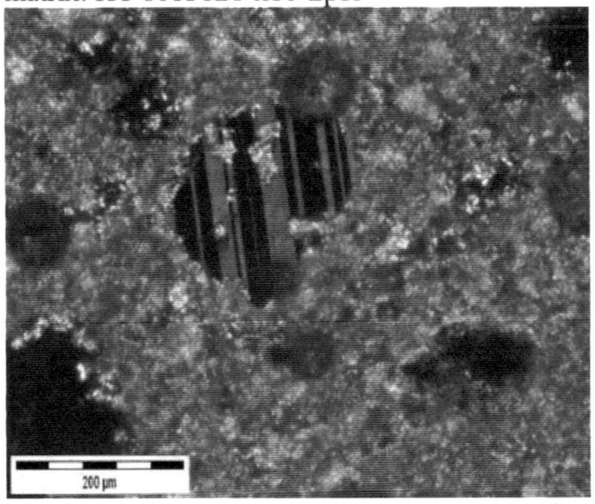

Figure 7. Plagioclase with albite composition in a calcareous matrix. HT80131160-x20-2pol

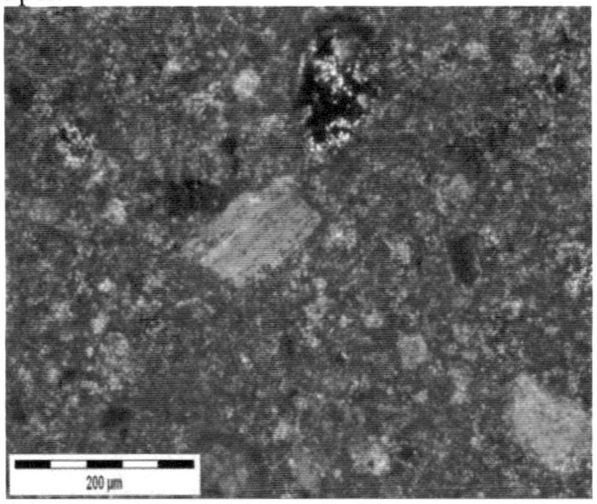

Figure 8. Biotite with typicaly cleavage on the surface and pleochroitic effect in a Ca-rich matrix. HT8013120-x20-2pol

4.4. Clustering and arrangement of data according to QXRD results

For classification and characterization of ceramics in a dendogram some parameters must be used for classifications. Firstly, a comparison of all analysis scans reveals a correlation matrix that demonstrating the similarity of scans. Secondly, agglomerative hierarchical cluster analysis puts the scans in different classes defined by their similarity (Bolshakova & Azuaje, 2003). The ancient ceramic materials are the objects that had their firing processes from a long time ago. In such a system the best phases are Quartz and Alkali feldspars, due to their stability in high temperature rang. In ancient ceramic system from Haft-Tappeh the sinter process is controlled by alkalifeldspars and quartz constituents.

The existence of diopside and alkalifeldspars in the matrix demonstrate the sinter temperature about 950°C < T < 1100°C, then the alkalifeldspar smelt by higher temperatures. The dendogram shows that the analysis divides itself into two different groups (Fig. 9). According to QXRD results the raw material is autochthonous from the region, but these two clusters show a transition on producing technology during 15th B.C. in Haft Tappeh.

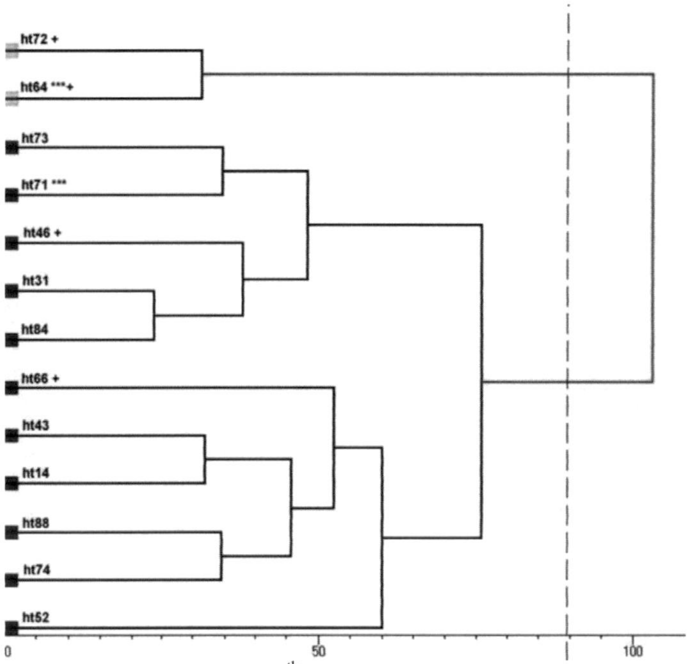

Figure 9. Cluster analysis of the samples from 14th B.C. from Haft-Tappeh. Dendogram interpreted according to the system Quartz – Albite – Orthoclase. The *** after a scan name indicate the most representative scan of a cluster. The + after a scan name indicates the two most different scans within one cluster

5. Conclusion

Mineralogical-chemical investigations with supplementary cluster-analysis on ceramics from Haft-Tappeh (15th B.C.) have given us useful interpretations about the manufacturing methods for pottery making in this area. The general chemical composition based on crystalline phases illustrate in system $SiO_2 - Al_2O_3 - (CaO+Na_2O+K_2O) - (MgO+\Sigma Fe_xO_y)$ (Fig. 10). The segregated part from tetraeder diagram obtains CaO-rich materials as calcareous raw material for manufacturing of ceramics. K_2O and CaO used as flux and sinter temperatures decreased according to the stability field of these phases. Hence the decrease in

temperature also reduces the crystallization of mullite. The sinter temperature is around 1000-1100°C. The common minerals paragenesis is: (i) Anorthite – Calcite – Diopside, (ii) Gehlenite – Augite – Diopside – Ca-plagioclase and (iii) Anorthite – Diopside.
According to the results obtained from clustering of the ceramic's constituents and thermal analysis, it is understandable that the raw material was mostly constant but manufacturing was different. The atmosphere of the kiln changed due to shifting of calcite decomposition peak range as well as recrystallization peaks by 800°C to 900°C. Crystallization of wollastonite, diopside and anorthite would interpret as manufacturing process under high temperatures but not so high that mullite can occur. Different ceramic technologies had been prepared in Haft-Tappeh by means of dissimilar firing conditions and in some cases by different admixtures as filler in the ceramic's body.

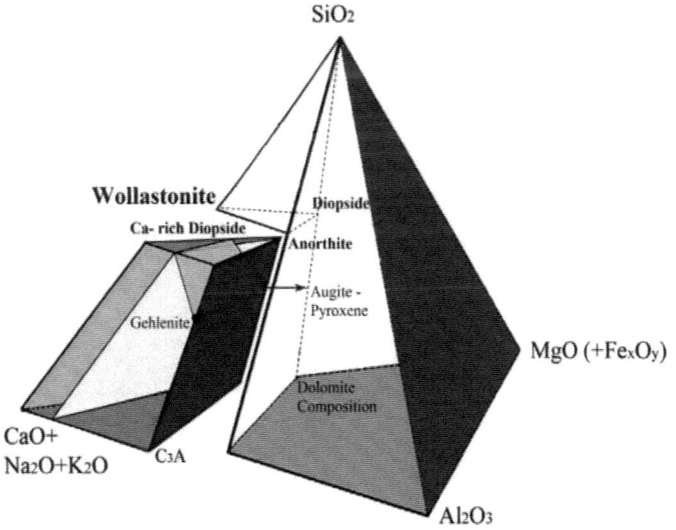

Figure 10. Mineral paragenesis in the ceramics from Haft-Tappeh in tetraeder diagram SiO_2 – Al_2O_3 – (CaO+Na_2O+K_2O) – (MgO+ΣFe_xO_y)

References

1. Bolshakova, N., & Azuaje, F. (2003). Improving expersion data mining through cluster validation. *4th Annual IEEE Conference on Information Technology Application*, (pp. pp. 19-22). UK.
2. Bullard, R. G. (1970). Geological Studies in Field Archaeology. *Biblical Archaeologist* , vol. 33, 98-132.
3. Emami, M. A., & Trettin, R. (2010). Phase generating processes in ancient ceramic matrices through microstructure investigation with high resolution microscopy methods. *Journal of Advanced Microscopy Research* , issue 5, vol. 11, pp. 181-189.
4. Emami, M., & Trettin, R. (2010). Phase Generating Processes in Ancient Ceramic Through Microstructure Investigation with High Resolution Microscopy Methods. *Journal of Advanced Microscopy Research* , vol. 5, pp. 1-9.
5. Emami, M., Kowald, T., & Trettin, R. (2009). Mineralogical and chemical investigation on the recrystallization process during sintering in face-interface areas in ancient ceramic matrices. *Materials and Manufacturing Process* , 24, pp. 934-941.
6. Emami, M., Volkmar, J., & Trettin, R. (2008). Quantitative characterization of damage mechanisms in ancient ceramic by quantitative X-ray powder diffraction, polarization microscopy, confocal laser scanning microscopy and non-contact mode atomic force microscopy. *Surface Enginnering,* , vol. 24, pp.129-137.

7. Gasparik, T. (1990). A thermodynamic model for the enstatite-diopside join. *American Mineralogist* , vol. 75, pp. 1080-1091.
8. Ghirshman, R. (1968). *Tchoga Zanbil (Dur Untash), Vol. II. Temenos, temples, palais, tombes* . Paris: Memories de la delegation archaeologique en Iran 40.
9. Heimann, R. B. (1978). Mineralogische Vorgaenge beim Brennen von Keramik und Aechaeothermometrie. *Acta Prehistorica et archaeologica* , vol. 9, issue 10, pp. 79-101.
10. Kambakhsh Fard, S. O. (2001). *Pottery in Iran, from Neilithic Period to the Present.* Tehran, Iran: Qoqnoos Publishing.
11. Magetti, M., & Gelatti, G. (1982). Die refernzgroupe(n) Lousana-mineralogische und chemische Untersuchungen der keramischen Produktion der Toepferwerkstaette Berna. *Jahresbericht der schweizerichen Gesellschaft fuer Ur- und Fruehgeschichte, vol. 65* , pp. 109-132.
12. Maggetti, M., & Schwab, H. (1982). Iron age fine pottery from Chatillon-s-Glane and the Heuneburg. *Archaeometry, vol. 24* , pp. 21-36.
13. Martineau, R., Walter-Simonnet, A. V., Grobety, B., & Bautier, M. (2007). Clay resources and technical choices for Neolithic pottery (Chalian, Jura, France), chemical. mineralogical and grain-size analysis . *Archaeometry, vol. 49* , pp. 23-52.
14. Matson, F. (1971). A study of temperatures using in firing ancient Mesopotamian pottery. In R. H. Brill, *Science and Archaeology* (pp. 65-79). M.I.T. Press, Cambridge.
15. Mofidi Nassrabadi, B. (2003). Archäologische Untersuchungen in Haft Tappeh. *Archaeologische Mitteilungen aus Iran und Turan* , vol. 35-36, Pp. 225-239.
16. Negahban, E. O. (1991). *Excavation at Haft Tape. Iran.* University Monograph 70, Philadelphia.
17. Noll, W. (1991). *Alte Keramiken und ihre Pigmente, Studien zu Material und Technologie.* D-7000 Stuttgart: E. Schweizerbart'sche Verlagsbuchhandlung .
18. Oudbashi, O., & Emami, M. (2010). a note on corrosion morphology on some midlle elamite copper aloy artefacts from Haft-Tappeh, south-west Iran. *Studies in Conservation, 55* , 20-25.
19. Heimann, R. B., Magetti, M. (1981), Experiments on simulated burial of calcareous terra sigillata (mineralogical change), British Museum Occasional Paper 19, pp. 163-177
20. Salmang, H., & Scholze, H. (1983). *Keramik, Teil 2. Keramische Werkstoffe.* 6. Auflage, Springer Verlag, Berlin, Heidelber, New York.
21. Stoeklin, J. (1968). Salt deposits of the middle east. In R. B. Edited Mattox, *International Conference on Saline Deposits,* Houston, Texas: Geological Society of America, vol. 88, pp. 158-181.
22. Tehrani, K. (1989). *Stratigraphy of Iran and Type Sections of Formations.* Tehran: Tehran University Publication.

IOP Conf. Series: Materials Science and Engineering **37** (2012) 012013 doi:10.1088/1757-899X/37/1/012013

Authenticity determination of Ag-Cu Lebanese coins using combined PIXE and RBS techniques

M Roumie, B Nsouli and Y Assafiri

Ion Beam Analysis Laboratory, Lebanese Atomic Energy Commission, National Council for Scientific Research, Beirut, Lebanon

E-mail: mroumie@cnrs.edu.lb

Abstract. A set of silver-copper coins were investigated for their stochiometry in order to check any possible counterfeit. The studied coins are 50 Lebanese Piastres and are relatively recent, minted in 1952. They are supposed to have nominal concentration of 60 wt.% Ag and 40 wt.% Cu. In this work, it is shown the utility and capability of combining PIXE (particle induced X-ray emission) and RBS (Rutherford backscattering spectrometry) techniques to determine the elemental content and check the presence, if any, of surface layer. For this aim, 3 MeV proton beam and Al funny filter, used as x-ray absorber, were used to determine the elemental composition, based on the emission and detection of the characteristic x-rays of the corresponding elements. The first PIXE results of the supposed bulky coins showed an excess of silver composition vs. copper. In addition, the experimental protocol was checked by analyzing some other coins of known composition, such as 1-euro and 2-euro coins.

1. Introduction

Many papers have already shown the strength of ion beam analysis techniques IBA, in particular PIXE, in the domain of investigation of ancient silver coins [1-3]. In fact, the precedent works focused on archeologist's problematic such as provenance, authenticity, counterfeit and fineness where, sometimes, an important financial stake is present. Therefore, IBA provided valuable information about the economy of the minting time, with the constraints that the analysis had to be totally non-destructive. Other X-ray analytical techniques, such as XRF and EDX which are less expensive than PIXE, could present similar or sometimes more advantages in determining the elemental composition, assuming the analyzed coin is mainly known as an alloy. However, the features of these X-ray techniques, including PIXE, are very limited and could affect the correctness of the outcome concentrations of the elemental structure, if the coin presents any surface layer. Thus, special attention was given, in such studies, on this particular issue, since evidence of silver surface enrichment of silver-copper coins was pointed out several times in the past few years. These layers could be due to wear, diffusion, corrosion or other effects [3-5]. Hence, the use of a combination of PIXE and RBS techniques would be very powerful tool [6,7], since RBS, unlike PIXE, could make a depth profiling of the compositional elements.

In the current work, it is shown the adaptability and capability of ion beam analysis, PIXE and RBS, to investigate the stochiometry of five contemporary silver-copper coins (Fig.1), in order to check any possible counterfeit. Nowadays, these Lebanese coins, minted in 1952, are available commercially on the market and present some significance among private collectors in many countries. In order to check the accuracy of our experimental protocol, some other coins of known composition were analyzed, such as 1-euro and 2-euro (Fig. 1), where the different seen elements were found to be Cu, Ag, Ni and Zn.

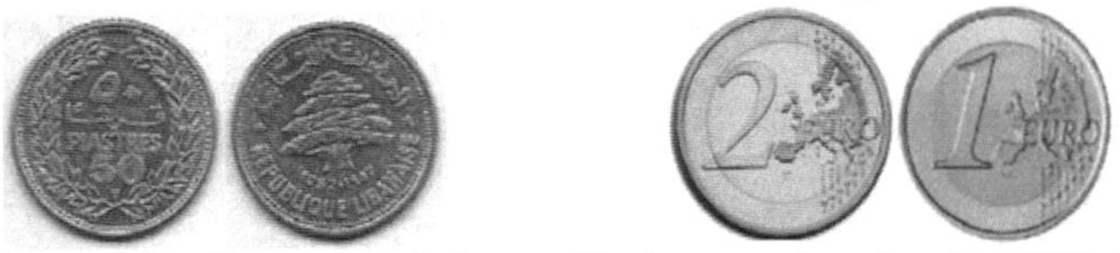

Fig.1: (left) the two faces of the silver-copper Lebanese 50 piastres coin, minted in 1952; (right) 1 and 2-euros coins were used as reference materials.

2. Experimental

The elemental composition of the samples was measured by the conventional in vacuum PIXE, using the accelerator facility at the Lebanese Atomic Energy Commission. It's a tandem Pelletron from NEC of 1.7 MV, model 5SDH, with an Alphatross RF ion source [8]. The multipurpose scattering chamber, 30 cm diameter, is designed to carry out PIXE, RBS and PIGE (proton induced gamma ray), where the beam hits the target under normal incidence. X-ray emission from targets is detected using an Ortec Si(Li) detector (SLP-06165 model) with 30 mm² active area, 12.7 µm thick Be window, and 170 eV measured FWHM energy resolution at 5.9 keV, placed at 135° to the beam direction.

In order to obtain in one spectrum most of the elements detectable by PIXE (low and high Z x-rays, majors, traces and probable impurities), the so-called funny filter, a pinhole filter drilled in its center, was employed as x-ray absorber [9]. A well calibration of the detection system (filter thickness, hole-diameter, solid angle or geometry) was done in order to have no energy dependence of the so-called H value [10]. An aluminum foil, 250 µm thick with a hole-diameter of about 0.3 mm, was found to present good detection sensitivity for most of the seen elements.

Simultaneously, RBS measurements were done under normal incident beam. A partially depleted PIPS (Passivated Implanted Planar Silicon) detector from Canberra, with 14 keV energy resolution and 25 mm^2 active area, detected the backscattered particles of the 3 MeV proton beam, at a scattering angle θ of 165° where the solid angle of the detector was 5.45×10^{-3} sr.

PIXE spectra were processed with the Gupixwin package (Guelph PIXE software package with Windows version) [11]. This software code is based on the fundamental parameter approach including x-ray production cross sections, x-ray attenuation coefficients, proton stopping powers, detector efficiency, collected charge and geometry effects to produce an output of elemental concentrations in ppm. RBS spectra were processed with the SIMNRA simulation code [12] to drive out quantitative thickness and structure composition.

Each of the studied five coins weighted 5.1 grams and had circular shape of 24 mm in diameter and 1.3 mm of thickness. Prior to analysis, no information was available on the surface enrichment or on the corroded surface thickness. Thus, they were slightly cleaned by diluted acid solution and then washed by distilled water, as corroded surfaces were well seen by naked eyes. After cleaning the shiny appearance supposed that the coins should be free of any corroded layer, however some authors attributed some surface enrichment to cleaning procedures [7]. Then, simultaneous PIXE and RBS analysis were done using 3 MeV proton beam and an Al funny filter in front of the Si(Li) detector. The beam current was kept lower than 2 nA in order to avoid high count-rate at the detector, which could reduce the detection sensitivity.

IOP Conf. Series: Materials Science and Engineering **37** (2012) 012013 doi:10.1088/1757-899X/37/1/012013

3. Results and discussion

To check the reliability of the experimental protocol, the known composition of the central part of the 1 and 2-euro coins, was chosen to validate our measurement on the five contemporary studied coins. These two coins (euros) were chosen due to the repeatability of PIXE analysis over the years, despite using different beam energies and different filters, used as x-ray absorber [13]. PIXE analysis of the euro-coins revealed good agreement between experimental and given values as shown in table 1, while no noticeable layer that could indicate a surface enrichment or corrosion was noticed in the corresponding RBS spectra.

However, an overestimation of silver concentration (80 wt.% on average instead of 60 %) was determined in all five coins according to PIXE results presented in table 1, where the coins were supposed to be homogeneously bulky with no surface layer. Thanks to RBS acquisition, used simultaneously with PIXE, where a clear silver surface enrichment is detected in all five coins. Furthermore, despite the chemical cleaning and shiny appearance of the coins, the presence of corroded surface is detected through carbon and oxygen signals (Fig. 2a).

Table 1: The elemental composition provided by PIXE measurements, as weight percent, for the 5 coins of 50 piastres, as well as for 1€ and 2€ used as reference materials.

	Ni	Zn	Cu	Ag
1 €*	25.1 (25)		74.9 (75)	
2 €*	5.0 (5)	19.6 (20)	75.4 (75)	
Coin no. 1			16.5	83.5
Coin no. 2			26.2	73.8
Coin no. 3			19.5	80.5
Coin no. 4			20.5	79.5
Coin no. 5			19.1	79.9
Coin no. 5, after abrasion			38	62

** Reference: (certified values in %)*

Indeed, the cross sectional yields of these two light elements are enhanced when using resonant energies such as the non-Rutherford 3 MeV protons [14,15]. Concerning silver enrichment, there are several heterogeneous layers at the surface, of different stochiometries and thicknesses, but all are rich in silver. Among RBS spectra, a well defined layer on the surface is seen on coin no. 5 (Fig. 2a), which is estimated to 3.5 µm in thickness and having 14 wt. % of copper and 86 wt. % of silver (the richest layer in silver). Moreover, all the subsequent layers seen in the RBS spectrum are still high in silver content, knowing that the projected range of 3 MeV protons in such alloy is roughly 35 µm, according to SRIM estimation [16]. In order to specify better the substrate composition, a mechanical removal of some tens of µm from the surface was necessary. Therefore, an abrasion was done on one of the coins (no. 5). The related RBS spectrum (Fig. 2b) indicates a common behavior of bulk material with no signal of surface layer (except a tiny carbon layer at the surface of ~25 nm that could be attributed to some vacuum residues). Accordingly, the PIXE analysis (Fig. 3) should be considered as reliable and are in perfect accordance with RBS results. The two simultaneous measurements, done by PIXE and RBS, show almost the nominal values of the 50 piastres coin (40 wt. % copper and 60 wt. % silver), as illustrated in table 1, and hence confirmed the authenticity of the studied coins.

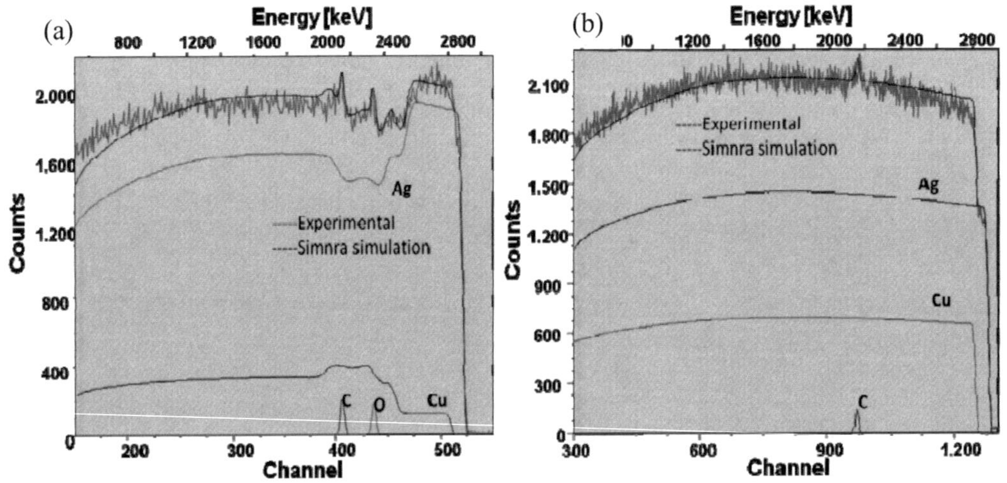

Fig.2: RBS spectra showing experimental, simnra simulation and elemental backscattered signals for the coin no. 5 before abrasion (a) and after abrasion (b).

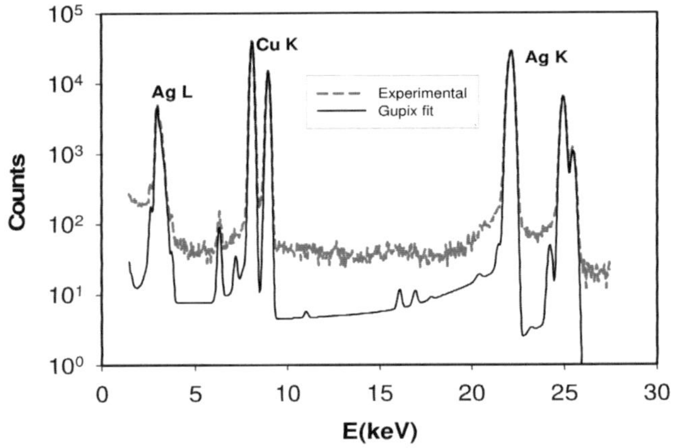

Fig. 3: PIXE experimental spectrum of coin no. 5 after abrasion along with the Gupixwin fitting. The resulted values are in good agreement with the nominal composition of the 50-piastres silver-copper coin of 1952.

4. Conclusion

This preliminary and concise study has highlighted the adaptability and capability of ion beam analysis to check the authenticity of silver-copper coins, minted in 1952. The proposed experimental protocol, based on simultaneous PIXE and RBS, was successfully applicable to the studied and reliable results were obtained. PIXE values of silver content were overestimated, since a silver enriched layer was detected by RBS acquisition. In addition, a thin corroded layer was observed trough the signals of carbon and oxygen which are undetectable by PIXE. However, PIXE perfectly quantified the homogeneous composition corresponding to the bulky coin, when RBS measurements showed no surface layers. The combination of PIXE and RBS techniques seems to be essential for an accurate measurement of silver coins composition, for authenticity and fineness studies.

Acknowledgments
The main author would like to thank *International Atomic Energy Agency* for their support of this project through the regional technical cooperation project no. RAS1011.

References

[1] Ben Abdelouahed H *et al* 2010 *Materials Characterization* **61** 59
[2] Šmit Ž and Šemrov A 2006 *Nucl. Instr. and Meth.* B **252** 290
[3] Meyer M A and Demortier G 1990 *Nucl. Instr. and Meth.* B **49** 300
[4] Linke R and Schreiner M 2000 *Mikrochim. Acta* **133** 165
[5] Klockenkamper R *et al* 1999 *Archaeometry* **41** 311
[6] Abraham M H *et al* 1999 *Nucl. Instr. and Meth.* B **150** 651
[7] Beck L et al 2008 *Nucl. Instr. and Meth.* B **266** 2320
[8] Roumié M *et al* 2004 *Nucl. Instr. and Meth.* B **219-220** 389
[9] Harrison J F and Eldred R A 1973 *Adv. X-ray Anal.* **17** 560
[10] Gama S *et al* 2001 *Nucl. Instr. and Meth.* B **181** 150
[11] Campbell J L *et al* 2010 *Nucl. Instr. and Meth.* B **268** 3356
[12] Mayer M 1997 *SIMNRA User's Guide, Report IPP 9/113*, (Garching, Germany: Max-Planck-Institut für Plasmaphysik)
[13] Roumie M *et al* 2010 *Nucl. Instr. and Meth.* B **268** 1916
[14] Gurbich A F 1997 *Nucl. Instr. and Meth.* B **129** 311
[15] Gurbich A F 1998 *Nucl. Instr. and Meth.* B **136-138** 60
[16] Ziegler J F 2004 *Nucl. Instr. and Meth.* B **219-220** 1027

Characterization of enameled glass excavated from Laem Pho, southern Thailand

W. Dhanmanonda[1], K. Won-in[2], S. Tancharakorn [3], W. Tantanuch[3], C.Thongleurm[4], T. Kamwanna[5], P. Dararutana[6]

[1]Faculty of Humanities and Social Sciences, Prince of Songkla University Pattani Campus, Pattani 94000 Thailand

[2]Department of Earth Sciences, Faculty of Science, Kasetsart University, Bangkok 10900 Thailand

[3]Synchrotron Light Research Institute, Nakhon Ratchasima, 30000 Thailand

[4]Science and Technology Research Institute of Chiang Mai University, Chiang Mai 50200 Thailand; Plasma and Beam Physics Research Facility, Department of Physics and Materials Science, Faculty of Science, Chiang Mai University, Chiang Mai 50200, Thailand

[5]Integrated Nanotechnology Research Center, Department of Physics, Faculty of Science, Khon Kaen University, Khon Kaen 40002 Thailand

[6]The Royal Thai Army Chemical Department, Bangkok 10900 Thailand

E-mail: pisutti@hotmail.com

Abstract. Laem Pho in Surat Thani, southern province of Thailand is one of the most important historic site on the eastern shore of the Gulf of Thailand. In this work, the enameled glass fragments which looked-like Islamic glass mainly excavated from this site were analyzed using SEM-EDS, PIXE and μ-XRF, in order to understand the chemical composition by comparing the archaeological data and topology. The structure of the enameled decoration was also studied. The resulting data indicated that high-magnesia alkali-lime silicate glass was produced. The presence of transition metals such as copper, iron and manganese were affected on the glass colorations. Typological classifications, technological observations and comparative studies serve to clarify the development and cultural inter-relationships of various glass objects along the trade and exchange networks in ancient maritime.

1. Introduction

The rise of the powerful force of Islam in Arabia by the end of the 7th century A.D. and its expansion, resulting in the political integration and broader Islamic acculturation of Iraq, Egypt, Syria, Iran, North and East Africa and southern Spain. The growth of great urban centers caused more demand for local and foreign commodities of all kinds, which included the goods from China and Southeast Asia. The archaeological evidences points to the commercial interaction between Middle East, Southeast Asia and China. China was the grand market and had the foreign colony at Canton, mostly composed of Persians and Arabs [1-2]. While Southeast Asia located on maritime trade route, many ports were situated at Malay Peninsula in order to serve the foreign merchants, such as the accommodations, trading spices and transporters including acting as agents, especially Laem Pho, a connection port between China and Middle East.

Laem Pho is an archaeological site located on the east coast of the Isthmus of Thailand, as shown in Figure 1. It is a sand spit, forming a peninsula that separates the Khlong (river) Yai Phum Riang from the waters of Bandon Bay. The river is wide and deep enough to form a secure anchorage for small ship; it may have been deeper in ancient times. This site has been visited by number of scholars; leading to several systematic excavations in the early 1980s. Particularly, between 1988 and 1989, the great excavation was carried out by the project so-called the Program for Ancient Settlement and Trade in Thailand. Many fragments of Chinese ceramics and Middle East wares were discovered [1-3]. The Chinese ceramics belong to the last Century of the Tang Dynasty (800-900 A.D.), especially Changsha bowl with Arabic character of Islamic name of God, Allah. Middle East wares were found to be both ceramics and glass forms. The ceramics which were found as glazed wares such as blue-green glazed ware, opaque white glazed ware, yellow enameled luster white ware, lead splashed ware and opaque white glazed with colored splash, were similar with those were found at Siraf (Iran), Mantai and Anuradhapura (Sri Lanka) [4]. The glass wares were mostly blue and green colored transparent and undecorated, except for few pieces which were decorated. These archaeological objects indicated that Laem Pho was an international port of trade on the east coast. It was a flourishing port during the 8th-10th centuries A.D. The abandonment of port has been interpreted as being related to the attacks by Rajendra Chola; King of South India who intended to destroy Srivijaya Kingdom. During this period, the sea level gradually rose so that shoreline advanced towards the port city. Laem pho was no longer the suitable the international port. Its function was therefore finished in the late 10th or early 11th century A.D.

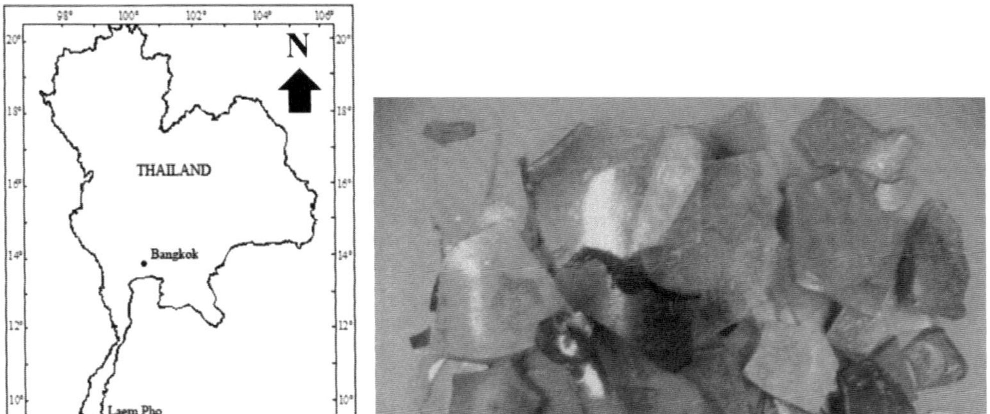

(a) (b)

Figure 1. Laem Pho Historic site; map (a) and glass fragments (b).

Decorated glass (luster painted or enameled) is a characteristic of the Islamic glass. It was painted with glassy pigments containing mostly silver and/or copper which fired in a kiln at a relatively low temperature of between 500 °C and 600 °C. During the firing, the silver metallic migrates through the surface into the external body of glass. Upon becoming reduces chemically to colloidal particles of metallic silver and becoming part of its atomic structure, it imparts a permanently coloration of the glass. The obtained colors depend primarily on the amount of silver present and the extent to which the chemical reduction had proceeded. It could range from a lemon yellow to strong amber. Evidently the presence of copper shifts the color further toward the amber shades. Today, the origin place of luster painting or enameled glass is still unclear, and is one of the subjects for discussion amongst historians. The excavation at Samarra (Iraq) revealed that the technique found to be satisfactory for production in the 9th century. It may have been developed in the early Abbasid period and overspread through Islamic political world [4-8].

In this work, the enameled glass fragment which was assumed as the Islamic glass that mainly excavated from this site were analyzed using scanning electron microscope coupled with energy dispersive X-ray fluorescence spectroscopy (SEM-EDS), proton-induced X-ray emission spectroscopy (PIXE) and micro-beam X-ray fluorescence spectroscopy (μ-XRF), in order to understand the chemical composition by comparing the archaeological data and topology.

2. Experimental

Glass fragment labelled LGLP4 that found at Laem Pho was characterized by using SEM-EDS, PIXE and μ-XRF. LGLP4, a blue glass body was decorated with polychrome colorants; amber and opaque light blue. Three areas were selected, as shown in Figure 2. They were marked as position#1, #2 and #3 for the amber colorant, glass body and opaque light blue colorant, respectively.

Due to their non-destructible nature, no sample preparation was allowed on these samples. Their structure and composition were analyzed using a Jeol 5410 LV SEM coupled with an Oxford Instruments INCA Penta FETx3 analytical system at the Institute of Product Quality and Standardization of Maejo University (Thailand). The system was operated at an accelerating voltage of 15 kV, an elapsed lifetime of 60s and a magnification of 500X. The chemical compositions were conventionally determined in oxide weight percent.

PIXE that was used to analyze based on a 2-MeV proton beam produced by a 1.7 MV tandem Tandetron accelerator at the Plasma and Beam Physics Research Facility of Chiang Mai University (Thailand). The proton beam was collimated with a diameter of 1 mm, and the beam current on the sample was 10 nA. The detector used was of Si(Li) type. Quantitative analysis of PIXE spectra of the chemical composition ($Z \geq 13$) in the samples performed using GUPIXWIN code. The quantitative calibration included the normalization at (100%-x) of the oxide sum, while the value of x is a sum of Na_2O and MgO determined by EDS.

Figure 2. Glass fragment LGLP4.

Prior to the analysis of the samples, their chemical composition measured using PIXE and EDS were compared. The standards used were optical glasses with known chemical compositions. It was found that the relative errors of the measured values for PIXE and EDS were less than 10 %. Results from PIXE, therefore, could be compared to those from EDS.

Micro-beam energy dispersive X-ray fluorescence spectroscopy (μ-XRF) using synchrotron radiation at BL6b of the Synchrotron Light Research Institute, Thailand was also used to confirm the presence of metals in the sample. The BL6b utilizes continuous synchrotron radiation from the bending magnet with energy range above 2 keV with the beam size of 100 μm at the sample position using polycapillary half-lens as an X-ray optics. To perform the experiment, the sample is placed on a three-degree of freedom motorized stage. Using high precision motorized stage allows us to accurately specify area or point of interest on the sample.

3. Results and Discussion

The compositional analysis of the sample obtained by PIXE and EDS measurements were illustrated in Table 1. In the glass body (position#2), major component was SiO_2. Na_2O showed relatively high concentration, while CaO concentration around 6 wt%. MgO is present with concentration ~3 wt%. K_2O showed high concentration approximately 5 wt%. Minor components determined were transition metals such as TiO_2, MnO, Fe_2O_3 and CuO, which correspond to impurities from raw materials used or intentionally added as colorants, especially in the case of MnO. The blue color was due to a very small cobalt content (0.1 wt%).

SEM investigation of the enameled glass surfaces (position #1 and #3) showed some microcracks from mechanical origin and small fractures because of surface weathering, as shown in Fig. 3. EDS microanalysis indicated that they were composed of a lead-based colorants. Major component was SiO_2, while the second main composition was PbO (up to ~15 wt%). The similarity of their main compositions was also found. The data required that the lead oxide were detected in both matrices. The enriched PbO may play also role of a network-forming oxide that made the lower melting temperature [9]. It was to create a white or bright yellow glass with thoroughly opaque that was as same as the production of lead-rich white glass in Roman times [10-11]. The amber enamel contained high concentration of copper oxide (~2 wt%). It was shown the present of silver (~1 wt%) in the opaque light blue enamel. The obtained colors were agreed with the previous works [4-8].

Table 1 Compositional analysis of the enameled glass sample, LGLP4

Oxide	Composition (wt%)		
	position#1	position#2	position#3
Na_2O	6.62	15.79	5.03
MgO	4.06	3.38	3.55
Al_2O_3	4.25	5.01	4.51
SiO_2	55.41	59.91	55.43
K_2O	4.73	4.92	4.58
CaO	5.02	5.69	4.80
TiO_2	0.06	0.39	0.03
MnO	0.95	0.99	0.91
Fe_2O_3	1.22	1.32	1.02
CoO	nd	0.10	nd
CuO	2.34	1.07	1.17
ZnO	1.01	0.85	0.53
SnO_2	0.22	0.30	1.28
PbO	14.02	0.68	15.12
Ag	nd	nd	1.11

Note: nd equals to non-detectable; <0.01 wt%.

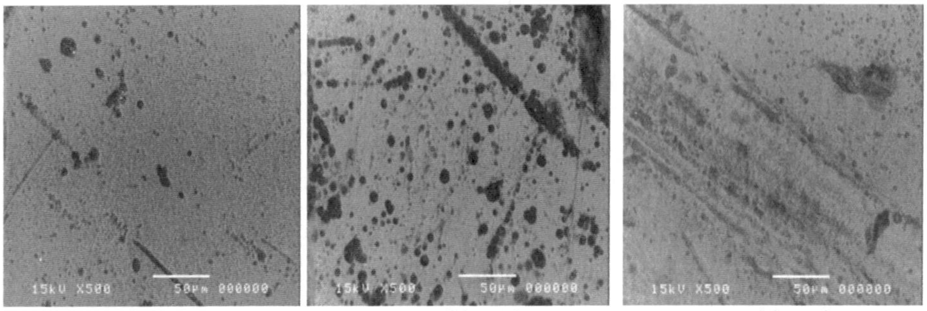

Figure 3. SEM micrographs.

The presence of metals on different positions of the sample has also been confirmed by micro-XRF techniques at BL6b. As mentioned earlier, specific area can be accurately chosen in order to compare elemental composition with different areas on the sample surface. However, due to the complication of quantitative analysis using continuous X-ray beam and limitation of light element detectable, only qualitative result will be described as shown in Figure 4.

The result from XRF reveals the elemental composition in the sample. Some metals including titanium, manganese, iron, cobalt, copper and zinc have been detected using this technique. It is also clearly seen that high concentration of copper has been found at #1 and #3. It indicates that copper is one the chemical composition in colorant. Noted that Ar(Ka) peaks shown in Figure 4 come from the sample environment.

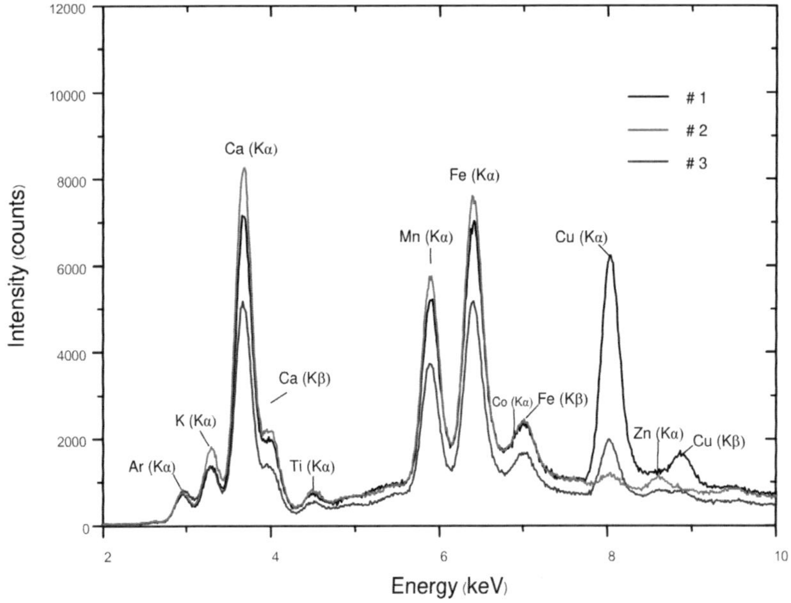

Figure 4. Micro-beam XRF spectra of the LGLP4 sample at different positions.

Results of the glass body identified as HMG glass that were agreed with the other authors that the magnesium content was useful for characterizing plant-ash glass produced in the Middle East and Islamic HMG glasses were produced between 8th and 14th centuries A.D. [12-14].

4. Conclusions

Up to now, the study showed the composition of the selected enameled glass fragment sample. The average composition of the glass body and enameled colorants obtained by PIXE and EDS measurements were based on alkali-lime and lead-alkali glasses, respectively.

SEM-EDS and PIXE that were non-destructive and non-sampling methods were proved in advantages to analyze the composition of the archaeological objects. Microscopy technique equipped with EDS microanalysis and PIXE provided chemical composition and morphology data, and trace elements, respectively. Also, the micro-beam XRF technique provided good-quality and quantity XRF for the elemental analysis with more advantages for samples with low mass density and small analysis areas.

Acknowledgements

Part of this work was the PhD Thesis of Faculty of Archaeology at Silpakorn University. The fund was partly supported by the Faculty of Humanities and Social Sciences at Prince of Songkla University Pattani Campus and the Faculty of Science at Kasetsart University. Authors are kindly thanks the 15th Regional Office of Fine Arts, Phuket (Major Boonyarit Chansuwan) for supporting the glass samples. The Plasma and Beam Physics Facility at the Faculty of Science at Chiang Mai University, the Institute of Product Quality and Standardization of Maejo University (Chiang Mai) and the Synchrotron Light Research Institute at Nakhon Ratchasima are thanked for providing PIXE, SEM-EDS and micro-beam XRF apparatus for analysis, respectively.

References

[1] Bronson B 1996 *Chinese and Middle Eastern Trade in Southern Thailand During the 9th Century A.D. Ancient Trade and Cultural Contacts in Southeast Asia* (Bangkok: Srimuang Printing Co.,Ltd.)
[2] Tampoe M 1989 *Maritime Trade between China and the West: An Archaeological Study of the Ceramics from Siraf (Persian Gulf), 8th to 15th centuries A.D.* (British Archaeological Reports 555: International Series) (Oxford: Archaeopress)
[3] Srisuchat A 1990 *Silpakorn Journal* **33** 26
[4] Seely P *et al* 1999 Glazed Ceramic Anuradhapura : The British-Sri Lankan Excavations at Anuradhapura Salgaha Watta: Volume 2 The Artefacts (Oxford: Archaeopress).
[5] Ettinghausen R 1942 *An Early Islamic Glass-Making Center: Record of the Museum of Historic Art* (New Jersey: Princeton University)
[6] Hirth F and Rockhill W W 1966 *Chau Ju-Kua: His Works on the Chinese and Arab Trade in the Twelfth and Thirteenth Centuries, entitled Chu-Fan-Chi* (New York: Paragon Book Reprint Corp.)
[7] Brill R H 2001 *Glass of the Sultans: Some Thoughts on the Chemistry and Technology of Islamic Glass* eds Carboni S and Whitehouse D (New York: The Metropolitan Museum of Art)
[8] Singh N K and Kirmani M Z 2005 *Encyclopedia of Islamic Science and Scientists* (New Delhi: Grobal Vision Publishing House)
[9] Carmona N *et al* 2009 *Journal of Cultural Heritage* **10** 439
[10] Sayre E V 1967 *Summary of the Brookhaven Program of analysis of ancient glass*. In: *Applications of Science in examinations of Works of Art* (Boston: Museum of Fine Arts)
[11] Henderson J 1985 *Oxford Journal of Archaeometry* **30** 77
[12] Sayre E V 1961 *Science* **133** 1824
[13] Henderson J 2002 *Acc. Chem. Res.* **35** 594
[14] Gueit E *et al* 2010 *Journal of Archaeological Science* **37** 1742

IOP Conf. Series: Materials Science and Engineering **37** (2012) 012015 doi:10.1088/1757-899X/37/1/012015

Corrosion of ancient glass beads found in Southern Thailand

K Won-in[1], Y Thongkam[2], S Intarasiri[3], T Kamwanna[4] and P Dararutana[5]
[1]Department of Earth Sciences, Faculty of Science, Kasetsart University, Bangkok 10900 Thailand
[2]Department of Archaeology, Faculty of Archaeology, Silpakorn University, Bangkok 10200 Thailand
[3]Plasma and Beam Physics Research Facility, Department of Physics and Materials Science, Faculty of Science, Chiang Mai University 50200 Thailand
[4]Integrated Nanotechnology Research Center, Department of Physics, Faculty of Science, Khon Kaen University, Khon Kaen 40002 Thailand
[5]The Royal Thai Army Chemical Department, Bangkok 10900 Thailand

E-mail: pisutti@hotmail.com

Abstract. Glass has been used as ornaments and decorations in Thailand for several hundred years. The archaeological resources suggested that the ancient glass beads excavated in southern Thailand were made more than 1300 years ago. Initial findings revealed that there were number of difference in shade between the glass beads of difference colors. Energy dispersive X-ray spectrometer (EDS) system attached with scanning electron microscope (SEM) and particle-induced X-ray emission spectroscopy (PIXE) were firstly used to study the surface corrosion of the samples. SEM micrographs showed more corroded and flaked microstructure. These were contributed to the interaction of both the ground water and its dissolved chemical compounds.

1. Introduction

Artificial glass is a material having a systematic history of more than 5000 years. Glass in the ancient world was typically made of at least three components; a source of silica or sand, a flux and a colorant [1]. Various colors in glasses are caused by inclusion of metal ions. These metals are usually transition elements in the periodic table, which absorb characteristic frequencies of visible region of the electromagnetic spectrum as a result of the d-d electronic transitions [2]. Cobalt and copper were commonly used in pre-industrial glass technology to produce blue coloration have been used to ascertain the origin of coloring, which induced different blue tonalities [3-5]. Glass is mostly found as beads in South and South-East Asia where the earliest evidence of glass comes from the northern part of India and dated from the end of the second millennium to the beginning of the first millennium B.C. [6-8]. Due to their beauty, affordability, transportability and durability, glass beads rapidly became popular as items of adornment and ceremonialism. Different kinds of beads were made and exchanged [9]. For many years, glass beads have been charming and attractive enough for both old and new collections in Thailand. It is the starting point for people who are interested in the origin of the glass beads and the progress of all human societies in the world. Many ancient glass beads were found at archaeological sites of southern provinces of Thailand, such as, Chumporn, Surat Thani, Pang-nga, Krabi, Nakhon Si Thammarat, Songkhla, and Pattani. Initials findings suggested that there are different shades among glass beads of different colors. Surat Thani Province is the largest southern province, on the eastern shore of the Gulf of Thailand. This area was inhabited in prehistoric times by Semang (a Negrito people) and Malayan (an Austronesin people) tribes. Founded in the 3rd century, the Srivijaya kingdom dominated the Malay Peninsula and much of the island of Java until the 13th century [10].

In this work, blue-colored ancient glass bead samples from Laem Pho Historic Site in Surat Thani Province were characterized for surface corrosion using scanning electron microscope equipped with microanalysis system and particle-induced X-ray emission spectroscopy.

2. Experimental

Three blue-colored ancient Thai glass bead samples were excavated from *Laem Pho* in Chiya District, Surat Thani Province. The glass bead samples are mostly monochrome in color and annular in shape. All three samples exhibited the same signs of weathering or devitrification, as seen in Figure 1.

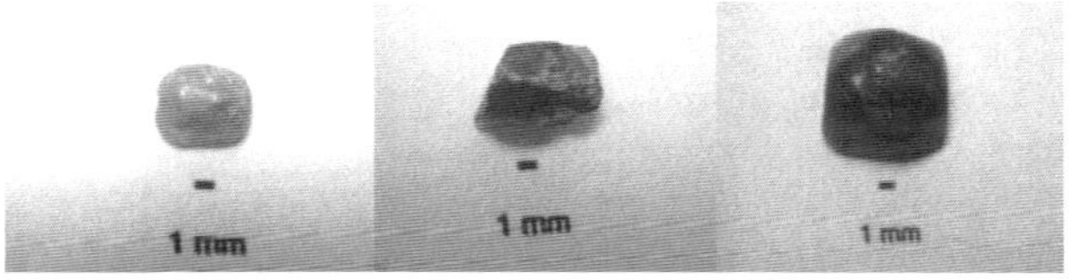

BPR2 BPR3 BPR4

Figure 1. Blue colored ancient Thai glass bead samples from Laem Pho, Surat Thani

Due to their non-destructible nature, no sample preparation was allowed on these samples. Samples were only cleaned with distilled water and then dried by hot air oven method for 24 hr to make them the same moisture. For each sample, three X-ray spectra were collected at different positions. Their structure and composition were analyzed using a Hitachi SU1500 SEM with a Horiba Emax energy dispersive X-ray fluorescence spectrometer (EDXRF) at Coax Group Corporation Co., Ltd. in Bangkok (Thailand).

PIXE that was used to analyze these samples based on a 2-MeV proton beam produced by a 1.7 MV tandem Tandetron accelerator at the Plasma and Beam Physics Research Facility of Chiang Mai University (Thailand). The proton beam was collimated with a diameter of 1 mm, and the beam current on the sample was 10 nA. The detector used was of Si(Li) type. A 74-µm mylar foil with 0.38% relative hole area was placed in front of the detector as an absorber to reduce the count rate caused by impurities of low atomic number elements. Quantitative analysis of PIXE spectra of the chemical composition ($Z \geq 13$) in the samples were performed using GUPIXWIN code. The quantitative calibration included the normalization at (100%-x) of the oxide sum, while the value of x is a sum of Na_2O and MgO determined by EDS.

Prior to the analysis of the samples, their chemical composition measured using PIXE and EDS were compared. The standards used were optical glasses with known chemical compositions. It was found that the relative errors of the measured values for PIXE and EDS were less than 10 %. Results from PIXE, therefore, could be compared to those from EDS.

3. Results and discussion

It was found that all three samples were transparent with a lot of bubbles. Table 1 contained analysis of the glass bead samples from *Laem Pho* site with calculated oxide weight percentages.

Laem Pho glass beads were classified as a low-magnesia soda-silicate glass (LMG) with SiO_2 content ranging from 74.57 to 75.81 %, Na_2O, CaO and K_2O concentrations ranging from 16.03 to 16.46, 5.55 to 5.63 and 5.19 to 5.37 %, respectively, and MgO content being below 1 %. They contained low magnesia and high alumina. Titanium oxide, a common mineral impurity in sand, was present with similar concentration in all samples. High ratio of the concentration of manganese oxide and ferric oxide in BPR2 indicated that manganese oxide was intentionally added as a decolorant [11].

BPR3 was sampling to study the weathering effect. Figure 2 showed SEM micrograph that revealed the surface detail of the weathered layer; more corroded and flaked microstructures were seen.

Table 1. Composition of the blue colored ancient Thai glass bead samples.

Sample	Composition (wt%)											
	SiO_2	Al_2O_3	CaO	Na_2O	K_2O	Fe_2O_3	MgO	TiO_2	CoO	CuO	MnO	PbO
BPR2	74.57	3.19	5.63	9.35	5.37	0.97	0.32	0.20	<0.01	0.03	0.89	<0.01
BPR3	75.81	3.76	5.57	9.03	5.19	3.93	0.42	0.29	<0.01	0.05	1.57	<0.01
BPR4	75.12	3.35	5.55	9.46	5.21	3.20	0.36	0.32	<0.01	0.17	1.08	<0.01
LOD(ppm)	2495	6166	178	-	138	78	-	26	121	41	22	1655

LOD equals to Limit of Detection of PIXE
Error of EDS equals to ±0.10 wt%

It was found that there were differences in concentration of composition at three different areas; area 1 was an original glass surface, area 2 was a corroded glass and area 3 was a new layer. The sample showed a decrease of alkali oxides (Na_2O and K_2O) and increase of SiO_2, detailed in Table 2.

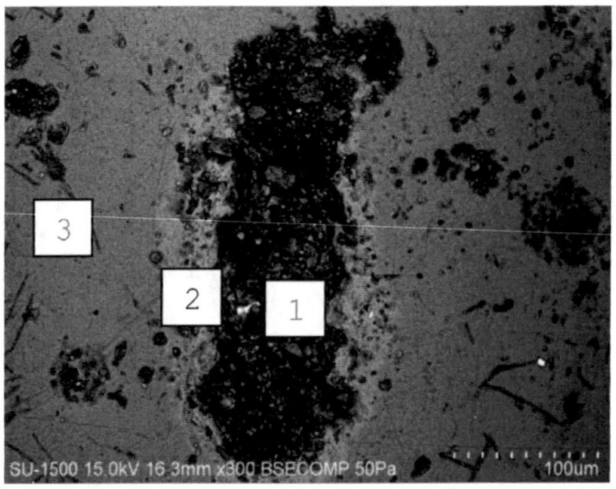

Figure 2. SEM micrograph obtained from a blue-colored glass bead BPR3; surface detail of the weathered layer.

Table 2. Composition at three different areas of BPR3.

Area	Concentration (wt%)		
	Na_2O	K_2O	SiO_2
1	10.652	6.827	72.024
2	3.882	4.280	72.862
3	2.320	3.781	82.691

It was known that the leading of alkali ions is a well-known weathered pattern in archaeological glasses [12-15]. The first step in the corrosion of glass involved an ion exchange of alkaline ions (Na^+, K^+) from the glass surface was due to H^+ ion from water, which promoted the formation of silica gel layer. The second stage was produced by the attack of hydroxyl ions (OH^-) freely available in the high alkaline aqueous medium, to the siloxane bonds (Si-O-Si) of the glass network. These bonds then were broken down to produce silanol groups (≡Si-OH) and non-bridging oxygen groups (≡Si-O$^-$), which implied gradual destruction of the glass network, giving rise to pits and craters. Under static condition, the reaction product enhanced the alkali attack to the glass surface, thereby producing a feedback corrosion mechanism.

IOP Conf. Series: Materials Science and Engineering **37** (2012) 012015 doi:10.1088/1757-899X/37/1/012015

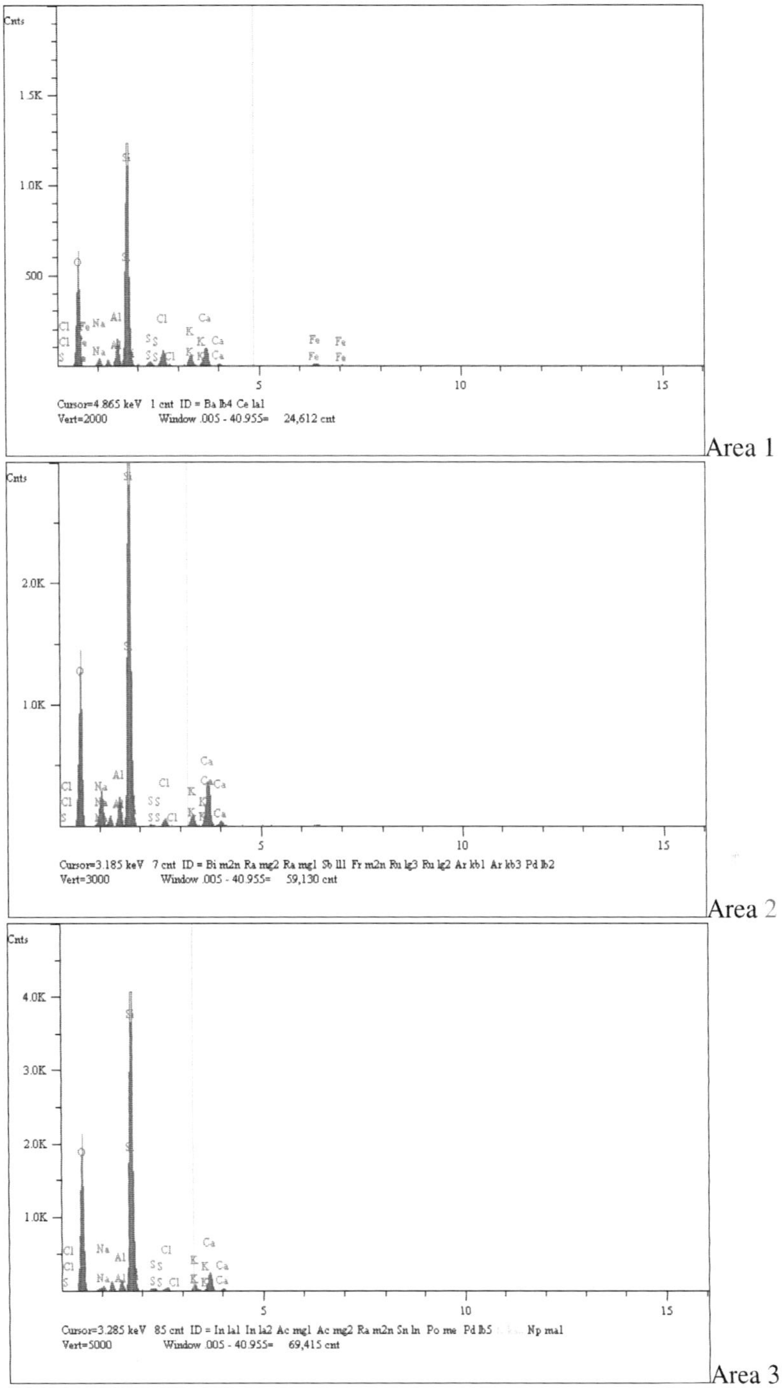

Figure 3. EDS spectra of BPR3

The high SiO_2 contents suggested that a protecting thin silica gel layer could be present on the surface of the sample. Such a layer could be originated as a result of hydration and acid attack during burial and could explain as well the loss by leaching of Na^+ and/or K^+ ions from the surface. On the contrary, the increase in SiO_2 was not detected in the sample with a more advanced alteration, even though in this case a higher decrease of Na_2O is readily appreciated. A feasible explanation for this

IOP Conf. Series: Materials Science and Engineering **37** (2012) 012015 doi:10.1088/1757-899X/37/1/012015

phenomenon could be that this glass was exposed to a stronger chemical attack due to its less stable composition. As corrosion proceeded following second and third stages, the firstly formed silica gel would be then destroyed, thereby producing a more intense dealkalinisation [16].

4. Conclusion

SEM and microanalysis of the blue-colored ancient Thai glass bead samples from two different sites in Surat Thani Province showed that they were both sodium- and potassium-silica glasses whose coloration was from copper and/or cobalt, which was influenced by the presence of iron and manganese. At the same site, slight compositional differences between glass samples may be a result of variables in glass production or the use of chemically similar raw materials with different batches. The poor glass quality and varying of identical composition may indicate a small scale production. Any results for light elements would have been unreliable because of the non-destructive analysis. The elements results may have been affected by not only the weathering but also surface leaching contamination.

Such corroded patterns were produced by the interaction of both ground water and its dissolved chemical compounds with the glass surface. In addition, water acidity also played an important role in the corrosion process of glass

The study of ancient glass, as a branch of archaeology, has centred on establishing the data and area of production of glass artifacts. Typological classifications, technological observations and comparative studies serve to clarify the development and cultural inter-relationships of various classes of glass objects.

Acknowledgements

This research was partly funded by the Kasetsart University Research and Development Institute and the Office of National Research Council of Thailand. The Plasma and Beam Physics Facility at the Faculty of Science of Chiang Mai University and Coax Group Corporation Co., Ltd. (Thailand) were thanked for their support on PIXE and SEM/EDS apparatus for analysis, respectively.

Referrences

[1] Shortland A J and Schroeder H 2008 *Archaeometry* **56** 947
[2] Nakai M *et al* 1997 *J. Phys IV France* **7** c2-1033
[3] Veiga J P and Figueiredo M O 2006 *Appl .Phys. A* **83** 547
[4] Frank S 1982 *Glass and Archaeology* (London: Academic Press)
[5] Henderson J 1985 *Oxford J. of Archaeology* **4** 267
[6] Eagle A 1976 *Readings in glass history; Glass in ancient India* (Jerusalem: Phoenix Publications)
[7] Francis Jr P 1984 *Journal of Glass Studies* **26** 152
[8] Kanungo A 2004 *Glass beads in ancient India: An ethnoarchaeological approach* British Reports, International Series 1242 (Oxford: John and Erica Hedges)
[9] Dussubieux L *et al* 2008 *Archaeometry* **50** 797
[10] www.wikimedia.org
[11] Hall M E and Yablonsky L 1998 *Journal of Archaeological Science* **25** 1239
[12] Garcia-Heras M *et al* 2005 *Journal of Archaeological Science* **32** 727
[13] Cox G A and Pollard A M 1997 *Archaeometry* **19** 45
[14] Freestone I C 2001 *Post-depositional changes in archaeological ceramics and glasses* Eds. Brothwell D R and Pollard A M Handbook of Archaeological Sciences (New York: John Wiley & Sons)
[15] Pollard A M and Meron C 1996 *Archaeological Chemistry* (Cambridge: The Royal Society of Chemistry)
[16] Fernandez Navarro J M and Vidrio E 2003 *Constitution, Fabrication, Properties* Third ed. (Madrid: CSIC)

IOP Conf. Series: Materials Science and Engineering **37** (2012) 012016 doi:10.1088/1757-899X/37/1/012016

Characteristics of ancient Egyptian glazed ceramic objects from Fatimid and Mamluk periods as revealed by ion beam analysis

Hamada Sadek [1 *] **Abd El Hady, M.M** [2]

1- Conservation dept., Faculty of archaeology, Fayoum University, Egypt.

2- Conservation dept., Faculty of archaeology, Cairo University, Egypt.

* E-mail: hsr00@fayoum.edu.eg

Abstract. Ion beam analysis (PIXE, μPIXE) has been successfully applied in analysis of archaeological materials, it has many advantages. In this work Ion Beam Analysis (IBA) used in analysis of ancient Egyptian glazed ceramic from 10th to the 16th centuries (Fatimid and Mamluk periods). Glazed ceramic samples from Al-Fustat Excavation store have been chosen to represent different colours (green, blue, brown, black …etc), the colours of glaze depend on many factors such as oxides present in the glaze layer, fluxes and the conditions in which objects had been manufactured in the past. Ion Beam allows the identification of elemental composition of the glaze layer i.e. ,the information about colorants used in glaze, which is of great importance for compositional data play a key role in solving questions concerning dating, provenance, technology, use and the relationship between ancient cultures with the environment.

1. Introduction

Ceramic was the first synthetic material created by humans, it was made up of a mixture of natural materials combined, formed into shape by a variety processes and transformed by heat to create a solid material not found in nature. Glazes are the coatings of the pottery surface and are silica-based the fusion of silica at low temperatures (700-800°C) is produced by adding flux agent such as lead or alkaline or both, these surface treatments can be applied to the dried pottery before or after the first firing and later subjected to subsequent heating treatment. Glaze charged with coloring agents gives a homogeneous color to the ceramic surface and, depending on the presence of opacifiers the glazed surface will be opaque or transparent. Decoration glaze also has a functional aim in protecting or impermeabilizing the pot's surface [Rice 1987].

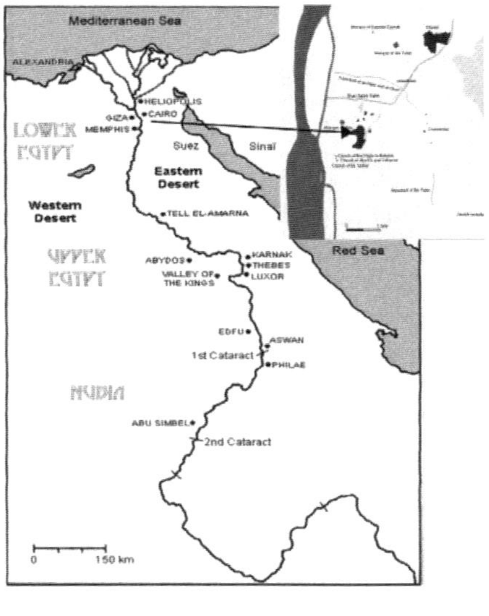

Fig.1 Map of Al-Fustat location near Cairo

In Islamic periods glazed ceramic was produced all over the Middle-East, however, one of its well known production center was Al-Fustat. Al-Fustat is the name of the first Islamic capital of Egypt, established shortly after the Islamic conquest of Egypt in 641A.C. It was built close to the Babylon fortress, which was a center of foreign marine trade because of its location on the Nile between lower and Upper Egypt map fig.1. Al-Fustat was an important center for ceramic production when Fatimid's and Mamluks overtook the Islamic Empire, it has played the role of the aesthetic stepsister, in spite of glazed ceramic manufacture was copious in Egypt its variety of shapes and decoration is impressive [Mason 1990]. Various investigations were done to identify the composition of ceramic objects, from mentioned archaeological site. In such researches the glaze was rarely taken into consideration , however, its composition is a source of valuable and supporting information [Hill 2004]. Compositional data may play a key role in solving questions concerning dating and provenance of ceramic objects, it can also provide with the knowledge about raw materials and colorant agents used during their manufacture and find out the details of glazes production technology. Glaze composition may also be an interesting source for information about the deterioration processes which often occur in archaeological samples as a result of the impact of the environmental conditions in opposite to many studies, which are focused on the glazes coloring technology.

The aim of presented research was to broaden the information about glaze coloration and fluxes technology used by craftsmen in Al-Fustat ceramic production center during Islamic periods the composition of glaze objects dated to this periods. Glaze compositional analysis requires employing the suitable analytical methods [Giussani 2009]. From year to year there can be observed the increasing role of Ion Beam Analysis (IBA) in the investigations of cultural heritage; it allows the direct analysis of solid samples without any previous preparation because of all mentioned advantages IBA was chosen as a method for presented investigation.

2. Experimental work:

2. 1 Sampling

Nine fragments of glazed ceramics were chosen for this study, all samples consist of a ceramic body covered with glaze layer which were decorated with various pigments giving different colors: black, yellow, brown, green, and blue. The samples are shards from different ceramic objects, for example: dishes, lamps, bowl and originate from one production area located in Al-Fustat tab. 1, samples refer to two different periods: first is Fatimid period (358-565 AH / 969-1169AD), second is Mamluk period (648-787AH /1250-1517AD).

Tab.1. Samples used in the study, collected from Al Fustat - Cairo.

No.	FF1	FF2	FF3	FF4	FF	FM1	FM2	FM3	FM4
Period	Fatimid	Fatimid	Fatimid	Fatimid	Fatimid	Mamluk	Mamluk	Mamluk	Mamluk
Color	Black	Brown	Green	Green	Yellow	Brown	Blue	Blue	Olive
Sample				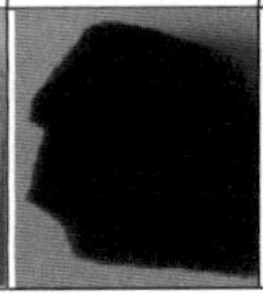			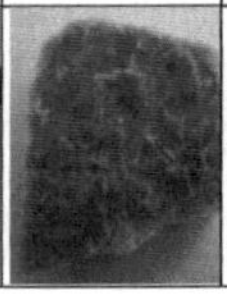	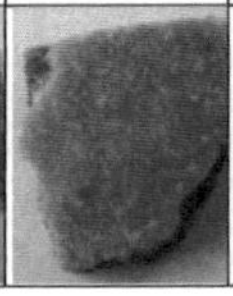	

2.2 Instrumentation

2. 2.1 PIXE Analysis

PIXE is one of the most sensitive micro analytical methods [Johansson 1995]. It is non-destructive and is currently used for multi elemental analysis both in vacuum and in air [Neelmeijer 2002], in vacuum mode can be advantageous for detection of low atomic number elements. Broad beam PIXE analysis with 1.0 MeV protons was performed at Warsaw – Poland using the Lech Van de Graff accelerator. Protons prior to entering the target chamber were collimated by two graphite collimator's with a diameter of 1 mm, the beam current was between 5 nA and 20 nA, the total charge during the

IOP Conf. Series: Materials Science and Engineering **37** (2012) 012016 doi:10.1088/1757-899X/37/1/012016

irradiation ranged from 0.1 μC to 0.5 mC. The X-rays generated in the samples were measured by a 7 mm^2 Amptek Inc. (Bedford, USA) Si (Li) detector with a 25 μm Be window, placed at 90° or 180° to the beam direction outside the experimental chamber and an energy resolution (FWHM) of 180–200 eV at 6.4 keV (checked during experimental runs by an ^{241}Am standard radioactive source). Samples with dimensions from 2x3 mm up to 10x10 mm were mounted on 8 position target holder perpendicular to the beam direction for analysis the final evaluation of data was performed for the spectrum resulting from subtraction of that taken for front side, this procedure made it possible to perform qualitative analysis even of glaze layer.

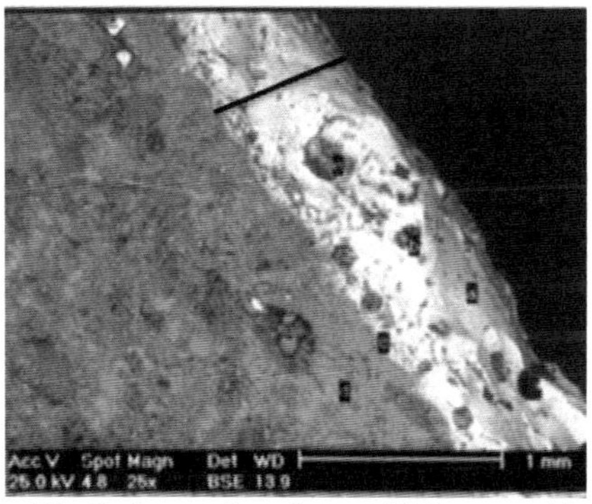

Fig.2 show glaze layer thickness- Al Fustat- Cairo

2.2.2 μ-PIXE analysis

The scanning micro beam-PIXE analysis allows identification of elements by taking high resolution elemental maps of the sample by scanning an area of 2 mm square with a narrow beam focused to 1 μm. A proton beam of 3.05 MeV from the 3 MV Tandetron accelerators at FZD (Forschungs zentrum Dresden, Germany) was used for μ-PIXE analysis. An Ortec Si (Li) detector was located outside the chamber behind a 125 μm thick are windows allowing detection of characteristic X-rays for elements heavier than Na. Since glaze layer thickness are usually of 0.1 –1 mm fig.2 [Abd el Aal 2009]. The analysis was performed in three steps: first the region of interest was selected with an optical microscope equipped with a CCD camera, next the beam of 5 x 5 μm size was scanned across the selected area usually 105 X 105 μm to get an elemental map, and finally in the region of maximum abundance of elements typical for a given glaze composition the μ-PIXE spectrum was measured with good statistics.

3. RESULTS AND DISCUSSION

The major and minor compositions of glaze samples were determined however, presented in this paper results and observations relate to major and minor elements content and its effect in glaze.

Black: Investigated sample are estimated as lead-alkaline glaze, however, with varying concentration of Pb and K. Mn was found as the main coloring agent, the results presented in fig.3, a Fe seems to play a role in the black coloration.

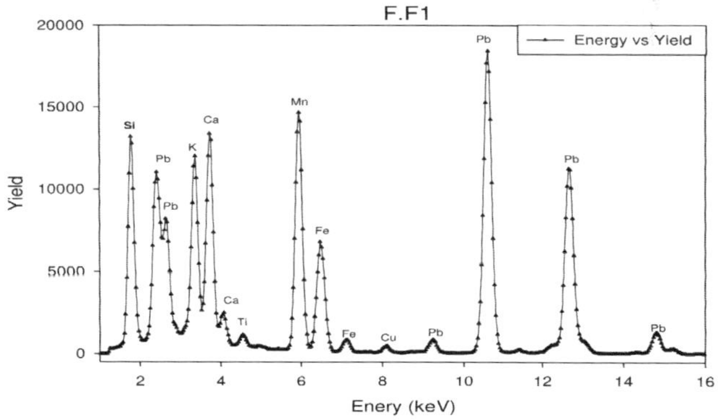

Fig.3 .Show μ-PIXE pattern of black glaze, Fatimid period Al Fustat-Cairo.

Brown: Among the two brown samples it was found that two of them are lead glazes, in which the potassium is found at trace level and it was detected by μ-PIXE .In both samples Fe acts as the main colorant agent, however, the hue varying from olive (FM4) to brown is an effect of different ratio in Fe, Mn, Cu concentrations in addition to firing atmosphere.

Yellow: In explored yellow sample group of glazes was found lead-alkaline glaze containing a high amount of Pb with mixture of K and Ca, yellow color seems to be an effect of the lead presence. In yellow glazes Fe appears as the minor contain. In the μ-PIXE map Sn identified as the opacifier agent fig.4, this may indicate that Lead-tin-antimony yellow was used as the colorant, prepared before the glaze manufacture, Although use of this coloring agent was not reported before for Egyptian glazes and is estimated to be first used in 18th and 19th century paintings in Italy, therefore this need to be confirmed upon the information about trace elements content and further investigations.

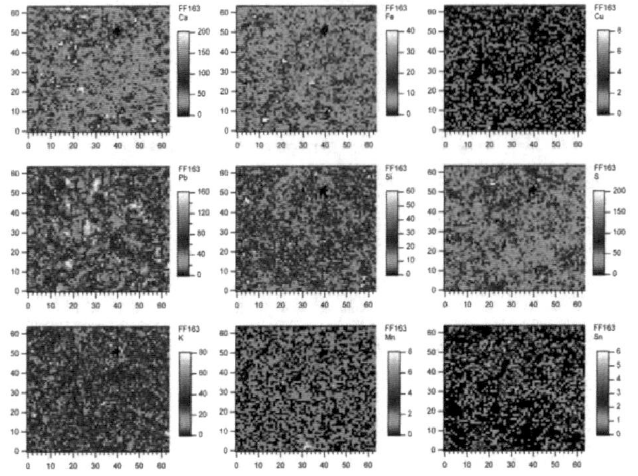

Fig. 4 Show μ-PXE map of yellow glaze sample- Al Fustat -Cairo

Green: both samples are lead-alkaline glaze, Cu and Fe with different concentration levels are responsible for the color in all types of green glazes, it is clear that with the use of raw materials with different ratio of those two colorants amount potters obtained different green glaze hues. However oxidation-reduction conditions played also a role in the matter of green hue obtained, therefore it is very difficult to estimate unambiguously the raw materials used in green glasses production [kotb 2005].

Blue: Two samples from Mamluk period were examined, Lead - alkaline oxides were determined as flux used in these samples, Colorant agent in the blue glaze is in most cases copper and its content depends on degree of blue color, Fe is the second element which with Cu plays a dominant role in hue changes. However for this element relationship with the content level and hue changes are not as direct as for Cu which indicates that correlation must be dependant also on the iron oxidation state in the glaze matrix. The more interesting fact is that during examining sample with the camera the blue hue similar to Egyptian Blue ($CaCuSi_4O_{10}$) was found. This pigment was used in ancient Egypt and there is no other evidence about using Egyptian Blue in Islamic glaze in Egypt. One of explanations may be the fact that technique of preparing Egyptian blue similar to way of preparing glaze fig.5 .Egyptian Blue was made by heating a mixture of sand, Natron ($Na_2CO_3 \cdot 10H_2O$) and a copper compound such as malachite. Pure silica has a melting point of 1700°C. Addition of flux, that role plays the sodium from Natron, reduces it to less than 1000°C and calcium oxide from limestone be included as a stabilizer [Lee 2000].

Fig.5. Blue glaze under microscope.

$$Cu_2(CO_3)(OH) + SiO_2 + 2CaCO_3 \xrightarrow[800\text{-}900^{\circ}C]{Flux} 2CaCuSi4O10 + 3CO_2 + H_2$$

4- CONCLUSION

Obtained results indicate that among these archaeological site glazes of different production technology is present. It could be mentioned also in one pigment of glaze we could sometimes estimate that there are more than one group of glaze color depending on the flux used during their production. The investigation demonstrates also that among one group of glazes, the samples of the same color may differ in the composition and this may have reflection in the different hues, however not in every case is visible to the first sight. This is an indication of different colorizing agent or technology used. Presented results in most of cases collaborate with previously established facts about colorization technologies applied in the past. Also some refer to not reported before colorizing agent usage appeared. Nevertheless colors of glaze depends on a lot of factors such as oxides present, fluxes added and the conditions in which objects have been manufactured in the past, therefore the presented statements and discussions are so far only hypothetic and need the confirmation of trace elements presenting in the samples. It could be concluded that IBA will be let's to obtain very interesting results and their interpretation will need also further studies let use to obtain very interesting results which interpretation still need to further studies.

Aknowledge

The authors gratefully acknowledge the Research Centre Rossendorf/Germany were supported by the European Community FMGE-CT98-0146 (Large Scale Facility AIM Rossendorf) Dr.Frans Munnik and Prof.A.Turos,A.Stonert from Soltan Institute of Nuclear Studies, Warsaw, Poland .

References

[1] Abd El Aal S. et al. (2009) ; Ion beam analysis of ancient Egyptian wall paintings ,Vacuum 83 S4–S8.

[2] Giussani B., Monticelli D., Rampazzi L. (2009); Role of laser ablation–inductively coupled plasma–mass spectrometry in cultural heritage research: A review Analytica Chimica Acta 635 6–21.

[3] HILL D. V. (2004) ;Chemical and Mineralogical Characterization of Sasanian and early Islamic Glazed ceramics from Deh Luran Plain,Southweastern Iran, Archaeometry 46, 4 585–605

[4] Johansson SA, Campbell JL, Malmqvist KG.(1995); Particle-induced X-ray emission Spectrometry (PIXE) . Wiley-Interscience.

[5] Kotb,H.(2005): Study of Technology, Treatment and Conservation of Archaeological Islamic Ceramic Lamps with Application on some Seleceted Objects, MSc thesis , faculty of archaeology, Cairo University.

[6] Lee,L. Quirke ,S. (2000); paintings materials in ; Ancient Egyptian materials and technology, By Paul T. Nicholson, Ian Shaw, cambridge university press .

[7] Mason , R . B & Edward, K .(1990) ; Petrography of Islamic pottery from Fustat , JARCE , PP . 165-183.

[8] Neelmeijer C, Mader M.(2002); Nucl Instrum Methods; B189:293–302.

[9] Rice, P.M. (1987): Pottery analysis: a sourcebook. University of Chicago Prcss, Chicago.

Understanding the Use-wears on Non-retouched Shells *Mytilus galloprovincialis. and Ruditapes decussatus* by Performing Wood Working Experiment: An Experimental Approach

Laxmi Tumung1, Behrouz Bazgir1, Kamran Ahmadi2 and Abdolkarim Shadmehr2
1 Department of Quaternary & Prehistory, Universitat Rovira I Virgili, Av. Catalunya 35, Tarragona ,Spain, 43002.
2 Material & Energy Research Center, Karaj, Iran.

ABSTRACT: This paper is an experimental attempt to understand the use-wear comes on non-retouched shells *Ruditapes decussatus* and *Mytilus galloprovincialis*. These species have been selected due to their variation in shape, size and edge type. In wood working experiment *Celtus australis* wood is used to perform the activities like scrapping and cutting wood. The ESEM results show the usewears in the form of linear marks, edge rounding, edge facture, polish and micro-pitting. Experiments also showed some macro-fractures.
Keywords: *Ruditapes decussates, Mytilus sp.*, Experiments, Use-Wear, ESEM.

Introduction

From the prehistoric times, coastal people have exploited shells for various purposes to survive, from food to tool making due to its local availability. The absences of stone and metal tools, lead to questions about the evolution of tool technologies and site exploitation by the hominids for millions of years. In that circumstance, we have to think of other alternative materials to be used like perishable tools such as wood or may be the local available materials like shells. There are number of sites around the world where shell tools have been reported.

Carl von Linné was the first who recognized some shell maidens as material evidence for the culture of ancient peoples. After that more scholars came with their piece of work on shells. Till 1970's, studies are mainly based on the shell maidens and to understand their role in the past societies and their culture (Bar-Yosef, 2002). There have been many studies based on the manufacture of shell ornaments and tools but a very few have made an attempt to understand the traces coming over shells after production and use. Since then many scholars tried to understand the functional analysis of shell tools by early hominid like making experiments to produce tools out of the shells by retouching or snapping the tool. Some scholars like Bonomo (2009); Dockall (1996); Douka (2010); Lucero and Donald (2005); Prywolnick (2003); O'day (2002); Light (2002); Barton and White (1993); Tyree (1998); Ritter et. al. (1995) and Reiger (1981) analyzed the archeological shells and have reported the presence of use-wear or residue on them. Toth and wood (1989) and Cleghorn (1977) tried to retouch the shells to understand the tool typology of shells. Toth and wood (1989) and Choi and Driwantoro (2006) tried to use the retouched shell knife and check the cut-marks produced by them on the bones. Some like Schmidt et.al. (2001) and Cuenca (2009 and 2010) tried to understand the use-wear on archaeological samples and even how to predict them by performing some experiments.

Therefore, current experiment is an attempt to answer these basic questions concerning micro-wear on the shell tools such as, the prime feasibility of micro-wear studies on shells smooth and granulated structure. Our studies show that the microscopic traces are present and identifiable on shell tools. Though the experimental programs was a very short one, yet it allowed us to observe some general trends and shows the wear features vary systematically according to variation in type of shell, contact material and use action.

Method and Material

For the experiment, we have taken certain variables to control the experiment program which are shown in the table 1. *Ruditapes decussatus* and *Mytilus galloprovincialis* shell species have been used for performing the experiment. Similar species documented from the Mesolithic and the early stages of Neolithic in the Cantabrian Sea, Northern Spain (Cuenca, 2009). They are selected under the criteria of edge shape, hardness, shell structure, and its local availability. *Mytilus galloprovincialis* and *Ruditapes decussatus* collected from the local fish market to have a wide variety of options to select those which are having less alteration on the edge due to the natural or man handling. All the shell species selected are having their own natural edge sharpness. Therefore, no retouching or edge modification has been done.

Ref. no.	Type of shells	Worked material	species	Delineation		Working angle	motion	action	Hand	time
				d-h	d-s					
MY04	*Mytulus galloprovincialis*	Stem of fresh wood	*Celtis australis*	Convex	straight	70°	Transverse bidirectional	Scraping wood	Right hand	10
CL02	*Ruditapes decussates*	Stem of fresh wood	*Celtis australis*	Convex	convex	90°	Transverse unidirectional	Scraping wood	Right hand	5
MY03	*Mytulus galloprovincialis*	Stem of fresh wood	*Celtis australis*	Convex	straight	90°	Longitudinal bidirectional	Cutting wood	Right hand	10
PE03	*Pecton maximus*	Stem of fresh wood	*Celtis australis*	Serrated	serrated	90°	Longitudinal bidirectional	Cutting wood	Right hand	10
CL01	*Ruditapes decussates*	Stem of fresh wood	*Celtis australis*	Convex	convex	90°	Longitudinal bidirectional	Cutting wood	Right hand	10

Table 1: Principal variables of wood working process

For the experiments, fresh wood (*Celtis australis*) have been used to understand the type of surface alteration caused by them. As the research is planning for small scale experiments, very few materials were used but have been documented in detail to understand surface modification and its development. We have performed transverse bi-directional work action for scraping action and longitudinal bi-directional for cutting action as shown in figure 1.

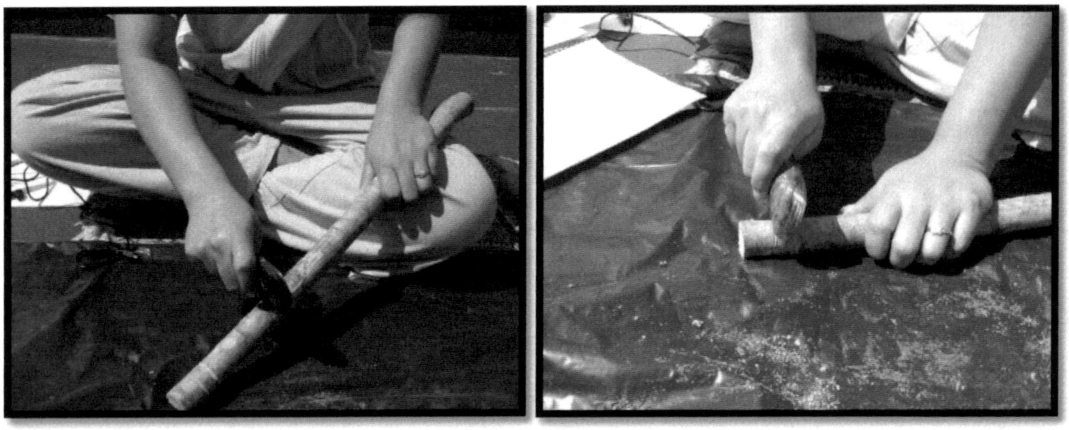

a) b)

Fig 1: a) Showing the transverse work action for scrapping the wood and b) Showing the longitudinal cutting action on *Celtus australis*

1. Cleaning process:

The cleaning procedures consisted of:

a) 10 minutes an ultrasonic bath in H2O2 to eliminate organic residues from shells before and after use,

b) 10 minutes an ultrasonic bath in the neutral phosphate-free detergent Derquim®, with ionic and non-ionic surface-active agents for, and

c) Rinse under cold running water to remove the extra neutral phosphate-free detergent Derquim® from the shell surface.

d) An ultrasonic bath in acetone for 2 minutes to eliminate the fatty residue resulting from the handling.

The steps (a, b and c) was used before and after the shell was used where as step d was performed before the microscopic analysis of the shells. After these various steps of cleaning the shells were and packed in different plastic bags to avoid future contamination and damages. This cleaning procedure has been shown to yield good results (Vergès, 2002; Ollé, 2003; Byrne *et al.*, 2006).

2. Preparation of Moulds:

Moulds and casts were prepared for the edges used to serve as reference for distinguishing micro and macro wear feature coming on the shells after use. Moulds were prepared with silicon based dental impression material, *Provil® novo Light* (Heraeus Kulzer, Inc.). The two components, a base and a catalyst in a ratio of 50% is taken on the impression material sheet and mixed under room temperature for 20-30 seconds so that it takes a uniform colour to prevent bad polymerization. Then the mixture was applied onto the shells using a spatula and left to dry. The moulds were not removed from the shells until they used for making cast to avoid any contamination inside the moulds.

3. Preparation of casts:

For preparing casts, a bicomponent rigid polyurethane resin, *Feropur PR-55* (Synthesia Española S.A.) was used. First a small amount of the mixture, mixed in equal proportion is poured in the mould with the help of thick opening needle syringe in the moulds as they were having very narrow opening for pouring the mixture in them. The moulds were kept standing still position and the syringe was moved inside the moulds so that the liquid penetrated the pores and then the rest of the mixture is poured rapidly as the resin starts hardening quickly. Casts have been considered very useful in sequential experiments as during successive uses edges can be damaged resulting in loss of control points. Casts allow us to observe fresh edges and add more control points at any stage of work (Vergès, 2002; Ollé, 2003; Ollé *et al.*, 2005).

Microscopic analysis

For the purpose of present study of usewear analysis on shell tools we investigated our samples under an ESEM (Environmental Scanning Electron Microscope) FEI QUANTA 600 from URV Scientific Resources Center. The samples were mound on a stub with the thermofusible glue in such a way that the edge of the sample is parallel to the stub. Before the ESEM observation, in order to find location, reference points were marked a little away from the tool edge with a 0.5 mm indelible felt tip pen and the same points were marked on a drawing of the specimen. The pen markings could easily be removed with a little bit of acetone. Then the tool was examined with the ESEM and observations were made on the microwear and images were taken. Then the cast studied in the same manner and then we analyze the same points for use-wear analysis that what changes have come on the used shells. The observations were made at a variable distances (from 10 mm to 30 mm) depending on the sample size in 20 kV. All the shell samples were analyzed in low vacuum pressure. For ESEM observations, three to four points of interest were located on each tool and then images were taken of each point of interest at several magnifications (ranging from 20x to 2000x). Higher magnifications were chosen to analyze the concrete detail whereas lower magnifications were used to study the distribution of the features.

Scraping experiment ESEM results

For this experiment, MY04 and CL02 were used. Due to the size of CL02 (3 cm), it was very difficult for prehension but was scraping quite well. After 5 minutes of use the sample broke into three triangle fragments as shown in figure 2b. As the sample is used for less time, very less use-wears occurred on the edge as shown in figure 2a. Just one edge fracture is observed which is having slight polish in the broken edge as shown in figure 3b. MY04 was good for scraping compare to CL02 and showed very good wears on the edge as shown in figure 3. *Mytilus galloprovincialis* first experienced the removal of periostracum and after that further use-wear developed on the exposed hypostracum. No striation marks have been observed in our experiment for wood working. It showed some polished on the edge whose texture was rough as shown in figure 3c and 3d.

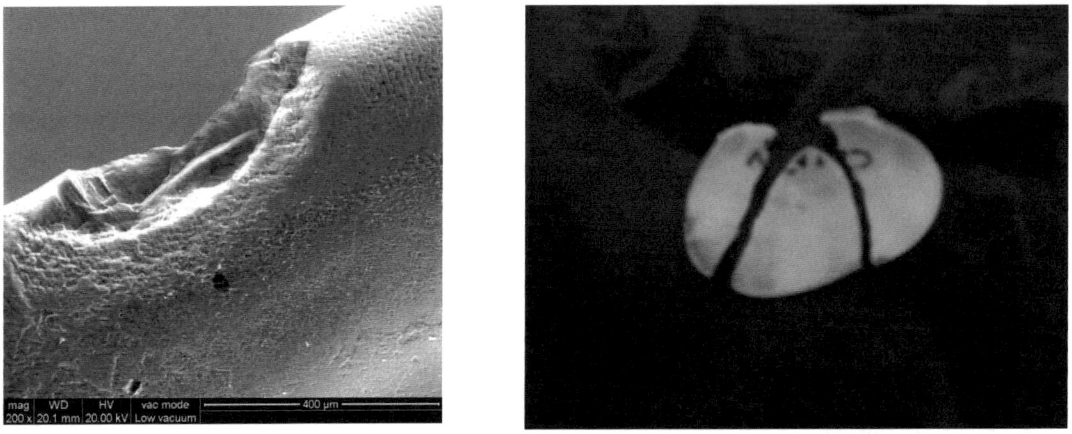

a b

Fig 2: a) Use-wears and edge fracture on CL02, b) broken into three fragments after 5 minutes of use.

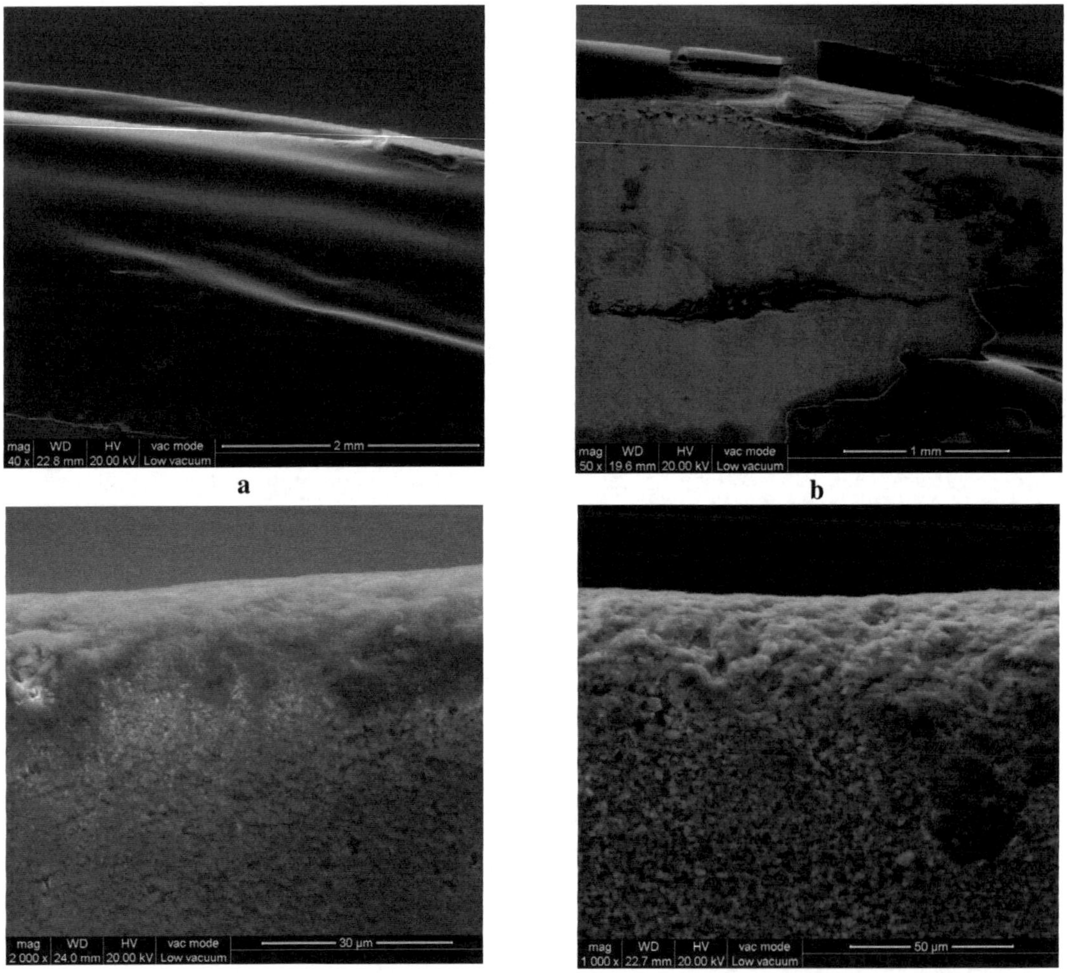

a b c d

Fig 3: Use-wear on MY04 a) cast of unused edge with periostracum over the edge, b) removal of periostracum after use, c) formation of polish in band, and d) highly developed polish on the working edge.

IOP Conf. Series: Materials Science and Engineering **37** (2012) 012017 doi:10.1088/1757-899X/37/1/012017

Cutting wood experiment ESEM results:

Same as MY04 sample which have been used for scraping action, MY03 also experienced the removal of periostracum and after that further use-wear developed on exposed hipostracum as shown in figure 4b and 4d. Very striation marks have been observed. They were deep and parallel to the edge. Edge rounding also has been observed throughout the edge. Edge also showed some micro-chipping on the surface of the polish and edge rounding. Polish in some places are in bands and parallel to the edge with maximum width of 500 mm. the edge which was frequent in use with the wood is having very smooth polish whereas other part which comes occasionally in contact with the wood are occurred in patches with some grooves on them. Edge fractures have been observed on many places.

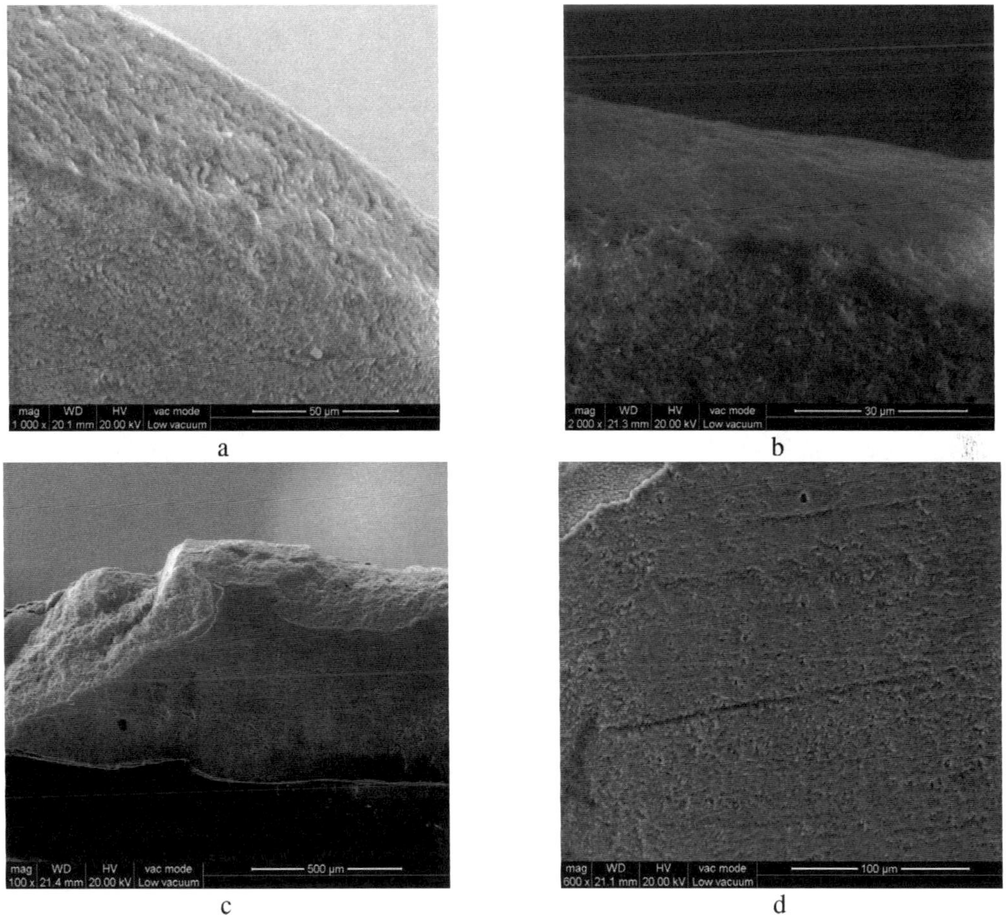

Fig4: a) Use-wears on MY03; (a and b) various types of polish and edge rounding after use, c) edge fracture and d) detail view of striation at higher magnification.

In the case of CL01 the edge before use had granulated surface and after the use surface slowly started to peel with the use as shown in figure 5a and 5b. Continues longitudinal bidirectional use of edge like saw for cutting the wood, make the granulated surface of the shell in an even plain. Edge rounding was present from before but with working of the edge it become more distinctive, Polishes are very developed and occur mostly on the working edge by making the granulated surface smooth. Textures of polishes are very smooth as shown in figure 5d. Mostly striations are continues, parallel to edge, arranged closely to each other. Texture of striation is very fine and shallow. They basically occurred on the polish of the edge as shown in figure 5d. Surprisingly no macro-fracture is observed even being small in size and delicate; it worked perfectly for the desired time. Number of edge fracture has been observed throughout the edge as shown in figure 5c.

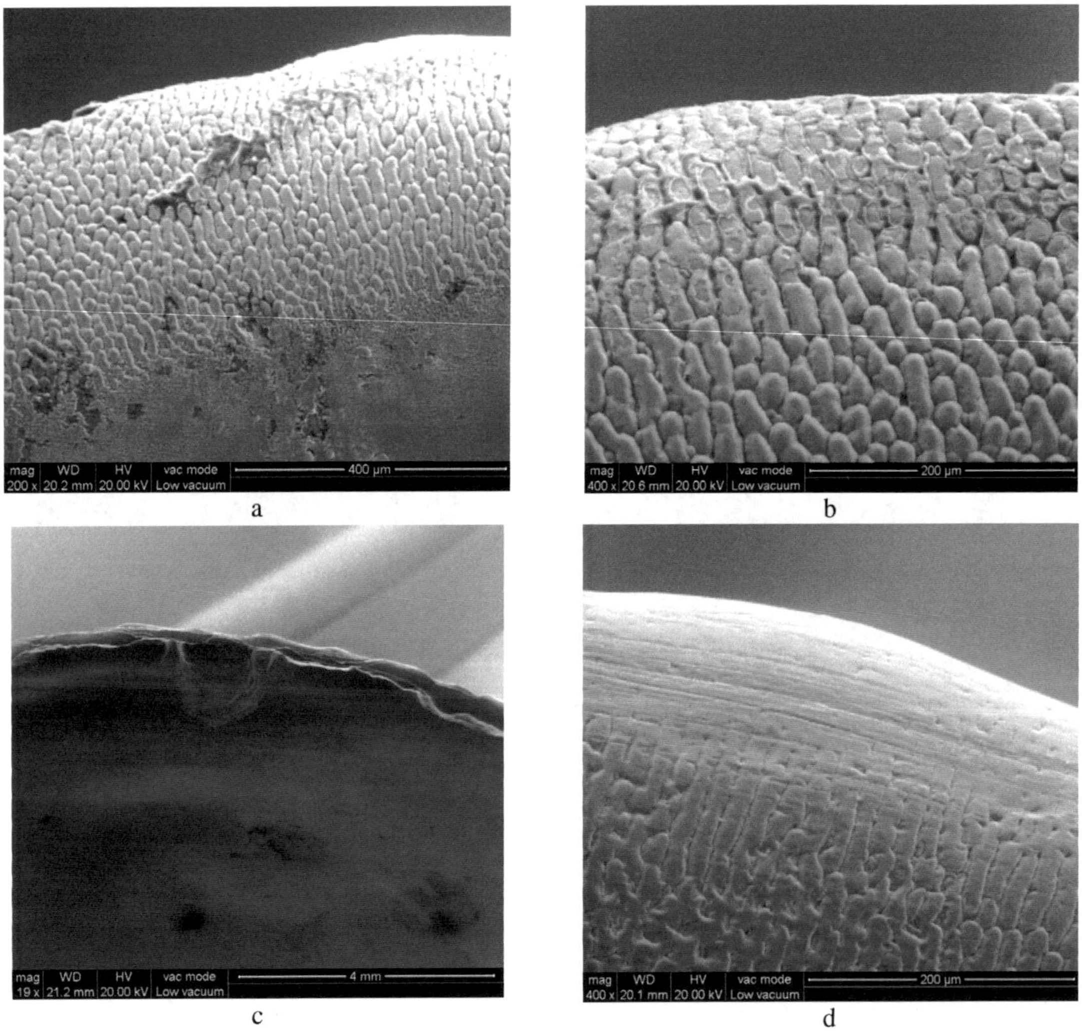

Fig 5: Use-wears on CL01; a) unused edge on cast, b) polish on granulated surface after use, c) edge fracture after use and d) edge rounding, polish and striation marks on them.

CONCLUSION

The experiments sought to answer only basic but important questions concerning micro-wear on shell tools. The prime question concerned the feasibility of micro-wear studies on different shell species. Our studies showed that microscopic traces of use-wear are definitely possible and yield fruitful results. As emphasized many times, this was not a wide experimental program devoted to explore in detail the traces by materials and by actions performed, but some major trends could be identified from our studies. The wear features included microfracturing, striations and linear features, polish, impact pits and edge rounding.

Micro-flake scars, striations and other linear features tended to be oriented in the direction of tool use. As the working edge was curve in all the cases, these features are parallel to the edge. Edge rounding was observed to be more intense on scraping tools and extending from the ventral onto the dorsal part due to greater contact with the worked material. Striations are more common in longitudinal action in cutting wood process but less in scraping action.

Polish has been considered to be one of the best indicators of contact material and distinctive of it on various tools. In our study trends in the appearance of polish showed that development of polish characteristics depends on contact material, use duration and useaction combined. Polishes are limited to higher relief areas in the micro-topography. Wood cutting samples showed more extensive distribution of polish than scraping of wood samples.

Macro-fracture also taken into account as it varies according to the type of shell used, its prehension and pressure applied to it while performing the experiments. Macro-fracture occurred most on the weaker shells species like *Ruditapes decussatus, Mytilus sp.* due to their shell size and strength.

This study is just a small step towards understanding micro-wear on various shell species. We need many more studies on this field in the future to understand wear formation on various shell species with contact to different materials in a better manner. Future studies would aim at widening the scope of the experiments to include more contact materials and use actions and undertake more experiments in each use material and use action category to understand the effect of different variables. Beside that as Biggs (1969) said, "Good cooperation between archeologist and malacologist is the only way to ensure the extraction of the fullest possible information on the ethnology of earlier population from shells". Therefore, such studies will help to improve the understanding of functional studies on shell tools and their role in human life from the prehistoric past.

Acknowledgement

We would like to thank Dr. Andreu Ollé Cañellas for his constructive suggestions and advices while doing this research. We are also thankful to the URV Scientific Resources Center for their cooperation and providing us all the facilities to carry out the Microscopic analysis.

References

Barton, H. and White, J. P. (1993). Use of Stone and Shell Artifacts from Bilof 2 , New Ireland, Papua New Guinea, *Asian Perspective* vol.32 no. 2, 170-183.

Bar-Yosef, D. (ed.) *9th ICAZ Conference, Durham 2002, Archeomalacology: Molluscs in Former Environments of Human Behaviour.*29-39.

Bonomo, M. & Aguirre, M. L. (2009). Holocene Molluscs from Archaeological Sites of the Pampean Region of Argentina: Approaches to Past Human Uses, *Geoarchaeology: An International Journal*, 24, (1): 59–85.

Byrne, L., Ollé, A. and Vergés, J. M. (2006) Under the Hammer: Residue Resulting from Production and Microwear on Experimental Stone Tools; *Archeometry,* 48, no.4, 549-564.

Choi, K., Driwantoro, D. (2006). Shell Tool Use by Early Members of Homo erectus in Sangiran, Central Java, Indonesia: Cut mark Evidence; *Journal of Archeological Science* vol. 34; 48-58.

Claassen, C. (1998). *Shells*. Cambridge, Cambridge University Press.

Cleghorn, P.L., (1977). A Note on Flaked Shell Implements: An Experimental Study, *Asian Perspectives,* vol. 20(2), 241-245.

Cuenca, D. S. (2009). Las "Tecnologias Invisibles" En los Groups de Cazadores Recolectores del Litoral Durante los Inicios del Holoceno (9.500-5.000 UNCAL BP) En la Region Cantabrica Utilizacion de las Conchas de Molusco en la Realizacion de Actividades Productivas; Master thesis; Universidad de Cantabrica DPTO of Ciencias Historicas.

Cuenca, D. S., I. Clemente-Conte, and I G Zugasti (2010). Using Shell Tools in Mesolithic and Early Neolithic Coastal Sites from Northern Spain: Experimental Program for Usewear Analysis in Malacological Materials. in *Trabajos de Prehistoria*, 67/1, 211-225.

Dockall, H. D. and Dockall, J. E. (1996). The Shell Assemblages from Morhiss (41VT1): An Archaic Site on the West Gulf Coastal Plain. *Southeastern Archaeology,* Vol. 15 (2), 211- 229)

Douka, K. (2010). An Upper Paleolithic Shell Scrapper from Ksar Akil (Lebanon), accepted manuscript in *Journal of Archaeological Science.*

Light, J. (2002). Marine Mussel Shells-Wear is the Evidence. In D. E. Bar-Yosef Maller (Ed) *Archeomalacology: Molluscs in Former Environments of Human Behaviour.*56-62.Oxford Oxbow Books.

Lucero J. M. and Donald J. S. (2005). Shell Tool in Early Holocene Contexts: Studies of Early Settlements in America Pacific Coast of Chile; *Current Research in the Pleistocene*, Vol. 22, 23-25.

O'Day, S. J. (2002). Late Prehistoric Lucayan Occupation and Subsistence on Middle Caicos Island, Northern West Indies. *Caribbean Journal of Science*, Vol. 38 no. 1-2, 1-10.

Ollé, A. and Vergés, J. M. (2003). SEM Functional Analysis and the Mechanism of Microwear Formation. In L. Longo and N. Skakun (eds.) *Prehistoric technology 40 years: functional studies and the Russian legacy*; BAR international series 1783.

IOP Conf. Series: Materials Science and Engineering **37** (2012) 012017 doi:10.1088/1757-899X/37/1/012017

Przywolnik, K. (2003) Shell Artefacts from Northern Cape Range Peninsula, Northwest Western Australia. *Australian Archaeology*; Vol 56, 12- 21.

Schmidt L., Atholl, A. and Fullagar, R. (2001). Shell and Bone Artefacts from the Emily Bay Settlement Site, Norfolk Island. In Atholl Anderson and Peter White *(eds) The Prehistoric Archaeology of Norfolk Island, Southwest Pacific*, 67–74. *Records of the Australian Museum, Supplement* 27. Sydney: Australian Museum.

Toth, N., and M, Woods. (1989). Molluscan Shell Knives and Experimental Cut-Marks on Bones. *Journal of Field Archaeology,* Vol.16. 250–255.

Tyree, K. D. (1998). Prehistoric Significance of Non-ornamental Modified Shell Implements from Baja California, Mexico. *Pacific Coast Archaeological Society Quarterly*, Vol. 34 no.3, 45-64.

Vergès, J. M. (2002). *Caracterització dels Models d'Instrumental Lític del Mode 1 a Partir de les Dades de l'anàlisi Funcional dels Conjunts Litotècnics d'aïn Hanech I El-Kherba (Algèria), Monte Poggiolo I Isernia La Pineta (Itàlia).* PhD Thesis. Departament d'història i Geografia. Tarragona, Universitat Rovira I Virgili.

IOP Conf. Series: Materials Science and Engineering **37** (2012) 012018 doi:10.1088/1757-899X/37/1/012018

Characterization of Some Iraqi Archaeological Samples Using IBA, Analytical X-ray and Other Complementary Techniques

Ziyad Shihab Al-Sarraj[1], **Mohamad Roumie**[2], **Hassan I. Damboos**[3]

[1, 3]Ministry of Science and Technology, Baghdad, Jadriya, P.O. Box; 765, Iraq
[2] CNRS, Beirut, Lebanon

ziyyad2004@yahoo.com[1], mroumie@cnrs.edu.lb[2], hassan_hdamboos@yahoo.com[3]

Abstract. The present work aimed at investigating the compositions and microstructures of some archaeological samples which dated back to various periods of the ancient Iraqi civilizations using PIXE, XRF, XRD, and SEM techniques. The models selected for the study (ceramics, glaze, etc.) were diverse in size and nature, therefore a limited number of samples were then butted from them by a small diamond wheel. Conventional powder metallurgy method was then used to prepare the samples. Dried samples were then coated with a thin layer of carbon, and analyzed using the ion beam accelerator of the LAEC. Three other groups of samples were also prepared for the purpose of analysis by X-ray fluorescence (XRF), X-ray diffraction (XRD), and scanning electron microscope (SEM). Analysis results of the chemical composition showed good agreement between the various techniques as well as for phases, while the fine structure analysis obtained by optical and scanning microscopy exhibited features of a structure where it got an intensified densification in the final stage of sintering and accompanied by quasi-homogeneous distribution of the closed pores. This will lead to the conclusion that the temperature used for sintering by ancient Iraqi was sufficient and it may fall in the range between 950-1200°C, also the mixes and the forming methods used by them, were both suitable to obtain good sintered bodies with even distribution of pores. A ring-shaped trace noticed in SEM micrographs need more work and study to explain what it is?

1. Introduction

Iraqi clays were well characterized by the presence of higher concentrations of lime and gypsum [1] in their compositions. Also it contains a smaller amounts of nitrogen and phosphorous (due to the lack of chemical decomposition), rock fragments, in addition to silt that comes with floods which may have a high concentration of metals. In general, southern Iraqi-clays were formed mostly from montmorillonite $[(Na, Ca)_{0.3}(Al, Mg)_2 Si_4O_{10} (OH)_2 . n (H_2O) \Rightarrow Na_{0.2} (Ca_{0.1} Al_2 Si_4O_{10} (OH)_2 (H_2O)_{10}$ which contains low percentages of aluminum [2]. All clays used in ancient times for Iraqi local potteries fabrication were of the types deposited by Tigris and Euphrates during floods. Over thousands of years huge amounts of potteries were fabricated in Iraq around the two tributaries of Tigris and Euphrates, a lot of which their fractured pieces were found dispersed over the surfaces of many ancient places related to all historic periods after 6000 BC. The endless number of pottery types all over the country of Chaldeans (Mesopotamia) considered of great importance for the determination of the age of the archaeological places in which they existed. The new and developed analytical techniques (physical, chemical, and nuclear) like ion beam analysis (IBA) using small accelerators in addition to the complementary ones like infrared (IR), X-Ray Diffraction analysis (XRD), and X-Ray fluorescence analysis (XRF) together with the various modifications [3-9] associated with them, provided the archaeologist, material scientists, and potters with the power to solve and answer all the questions related to the fabrication route, origin of raw materials, type of pottery and the patterns of surface treatments. Judgments or decisions about the precise historic period and the origin of pieces related to ancient potteries could be of high benefit for the archaeologists in the present days which will not be the case for those who already using potteries. The interest in the pottery's functions but not in their shapes and inscriptions considered to be familiar in the archaeology of Mesopotamia [10-11]. Among the regions of western Asia, Mesopotamia is unique in procession of a wealth of vocabulary for the names of pottery, which are often related to unique patterns (similar to vases and jugs) related to a ritual-workshops and not to the general pottery used in daily functions.

In the current work multiple types of potteries which were diverse in size, nature, ownership, in addition to the time periods and different civilizations were used. Specific models chosen were for the purpose of the present study which was focused on the analysis of their constituents and understanding the metallurgical structure in order to predict some of the conditions of manufacture.

2. Samples selection and preparation

In the present work different groups of archaeological samples were obtained from the general directorate of Iraqi national museums. A variety of samples were selected which dates back to various periods and civilizations such as: ancient Sumerian, Acadian and Babylonian, Assyrian modern, Verthy, Islamic, etc. Also the chosen models were diverse in size and nature, including parts of ceramic pots, cups or a broken piece of ceramic glaze or plaster. According to the nature of the test that will be done later, a limited number of samples were selected and as small as possible pieces were sliced by a small laboratory diamond wheel. Four groups of the samples were prepared for testing.

Steps used to prepare the samples and the related models selected are shown in Fig.1 (a, b).

2.1. First group, which composed approximately of 12-14 samples, prepared in the laboratories of Lebanese Atomic Energy Commission via conventional powder technology route for the sake of quantitatively analyzing their constituents using the ion beam (IBA) generated by the small nuclear particle accelerator available at CNRS, Beirut. Fig. 2 shows some samples before coating with carbon.

2.2. Second and third groups of the samples used for XRF and XRD analysis were prepared in the laboratories of the Iraqi ministry of science and technology (MOST). Ten samples of each group were prepared by a similar method used before excluding the carbon coating steps. All tests were conducted in Iraq.

2.3. Fourth group of samples for metallurgical investigations were prepared first in MOST laboratories, and then coated with a thin layer of gold for the suppression of static charge accumulation at the surfaces of the samples. Scanning electron microscope (SEM), existed in the centre of strategic studies at sulaymaniya (Kurdistan-Iraq), was used for the study of micro-structural features.

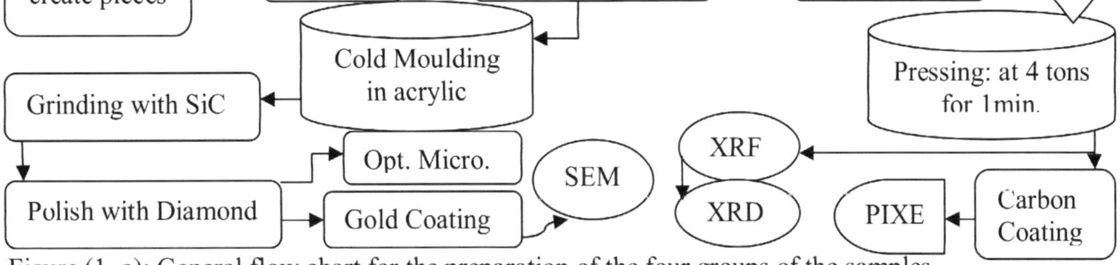

Figure (1, a): General flow chart for the preparation of the four groups of the samples.

IOP Conf. Series: Materials Science and Engineering **37** (2012) 012018 doi:10.1088/1757-899X/37/1/012018

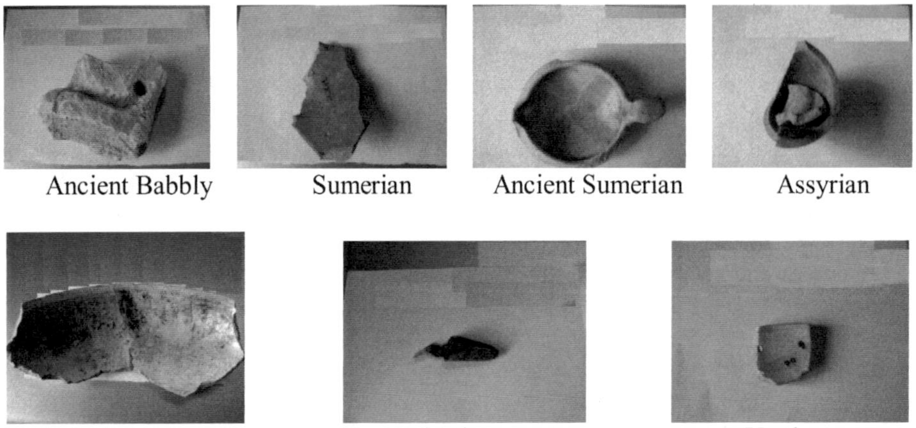

Figure (1, b): The real view of the selected models from which the samples were taken.

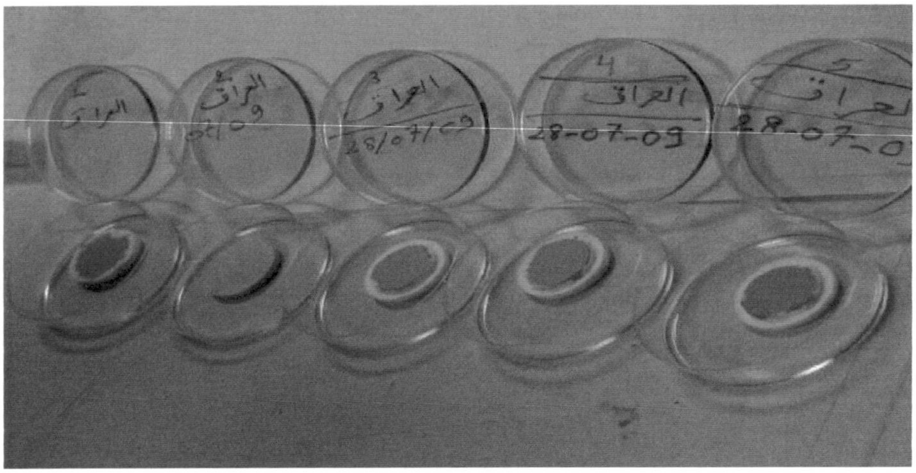

Figure (2): Shapes of some of the uncoated samples after pressing and confinement in an inert powder.

3. Results and discussion

For the archaeological samples under investigation, XRD patterns revealed the presence of phases which common for clay materials. Accordingly the study was then devoted to the determination of elemental constituents of the samples using XRF and ion beam techniques; Fig (3) shows an example of it. A comparison analysis of the data obtained reveals a reasonable harmony between the results of the two techniques when the concentration of the constituent elements was considered. For example ion beam analysis (i.e. PIXE) of ancient Sumerian (somery-1) revealed that silica (SiO_2) was the predominant phase followed by Calicia (CaO), Alumina (Al_2O_3) and finally iron oxide (Fe_2O_3). It was interesting to notice that approximately similar trends were exhibited by x-ray fluorescence analysis as shown in Fig (4). But when only the results of ion beam analysis were compared among the samples, it was noticed that the Acadian (Acady-2) sample processes higher percentage of Calicia (CaO). As compared to ancient Sumerian (somery-1) in addition to that; the percentage of (Fe_2O_3) and (Al_2O_3) was lower than that of summery. For Ancient Babylon (bably-3) sample processes higher concentration of (CaO) as compared with Sumerian Samples, although it was lower when compared

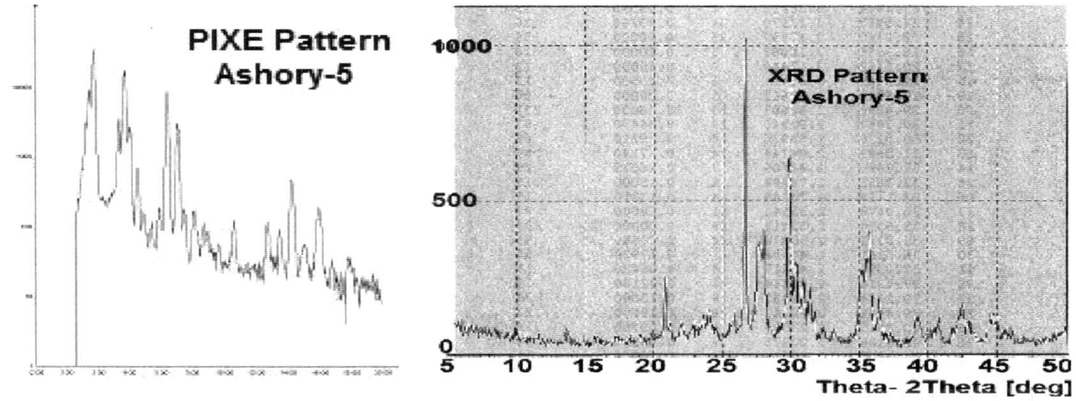

Figure (3): An example of the PIXE and XRD patterns obtained from testing by ion beam analysis (IBA) and X-Ray diffraction technique respectively.

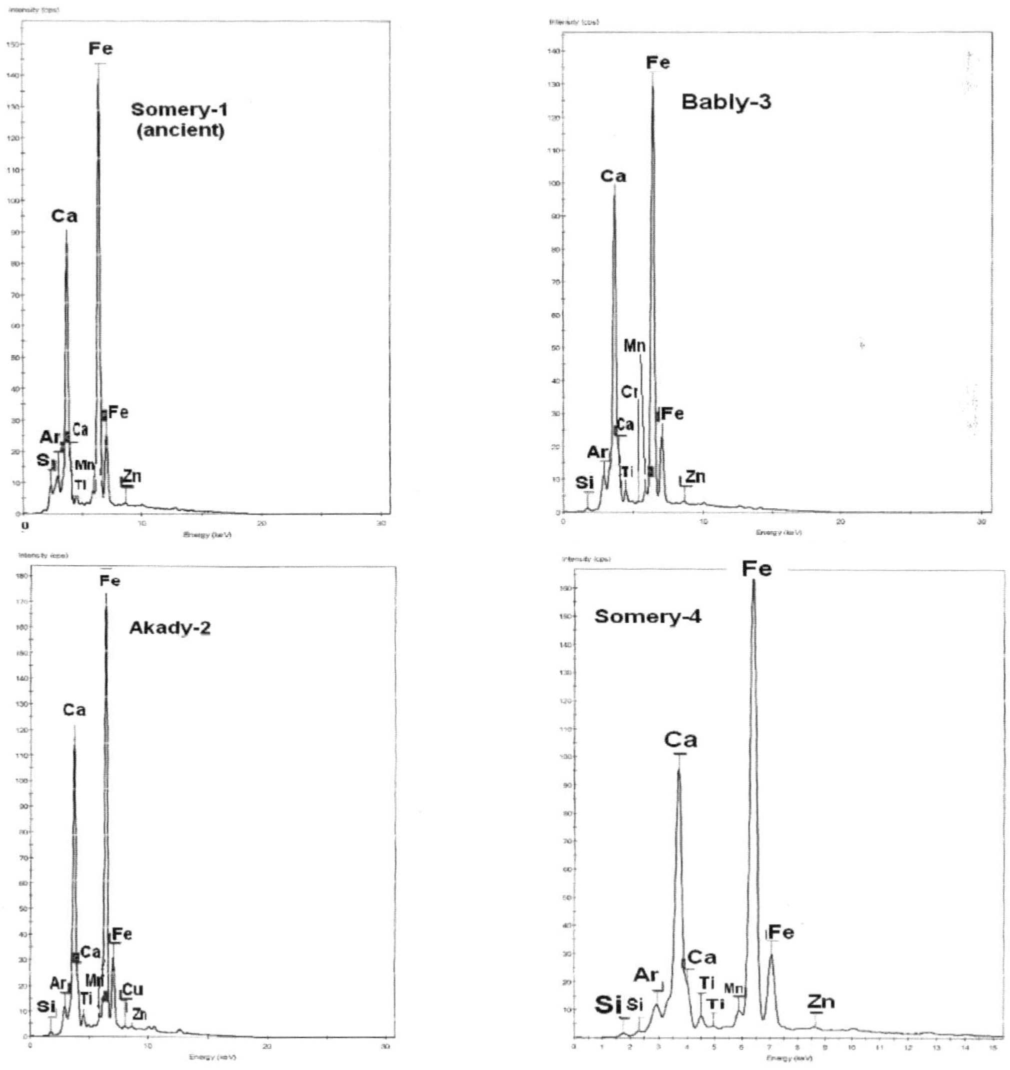

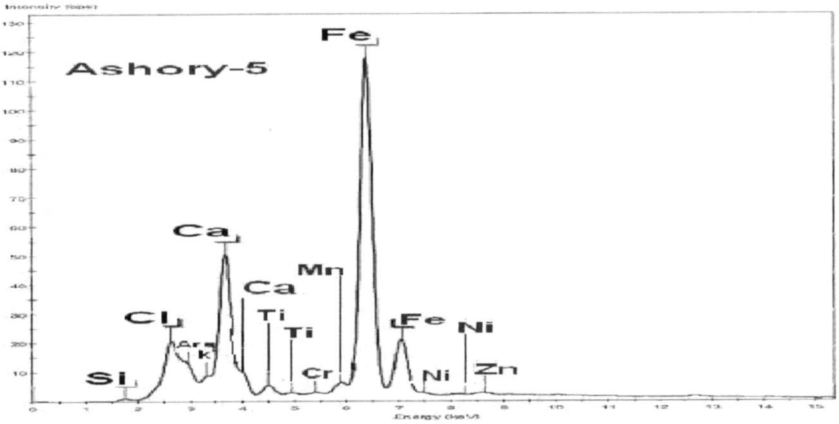

Figure (4): Show the XRF analysis of a five selected samples as pointed on each of them.

with that of (Akady-2) sample. Also, it was found that the Assyrian modern (Ashory-5) sample has the higher percentage Al_2O_3 which was followed by somery-4, Bably-3, and Acady-2. Finally the content of (MgO) was found to be somewhat convergent, and to be the highest in Assyrian sample (Ashory-5) and approximately the same in (Somery-4) and (Akady-2) sample and to be the lowest in (Bably-3) sample. For the sake of understanding and predict some of the variable parameters of the preparation method already used in the manufacturing of the archaeological articles under investigation, SEM was used to follow the microstructure and the densification features which reflects the preparation and , manufacturing parameters , in addition to the thermal history of the raw materials used. In Figures (5) of SEM-micrographs, the existence of spheroiding closed pores with homogeneous sizes in the range of 250-300 μm and its even distribution was clear. Any interconnection or channelling between the pores was not observed. Also the analyzed SEM micrographs were completely free from open porosity. Therefore the exhibited microstructure and its metallurgical features led to the conclusion that most of the studied potteries expected to be sintered at enough and suitable temperatures to attain such a microstructure, which in turns means that its fabrication conditions were properly selected by professional workers whom had good experience and so familiar with such methods, and all the work was done in working plants. Moreover the micro-structural homogeneity, even distribution of pores and the absence of large isolated clustered, merged or coalescences pores; clearly suggest that the fabrication method and design were suitable and neither random nor primitive but it could be considered as an indicator for the existence of tools and specialized workshops for ceramics and pottery articles. Of considerable interest, what was noticed for the Acadian sample in which a characteristic mark in the form of ring-shaped was stamped in the surface of the sample and completely appeared as a part of the sample surface since the pores existed entirely within it as for the rest of sample body. For the moment there was no definite explanation for this, and may be it represents a sign, a mark, or the manufacturers or most probable the fabrication methods.

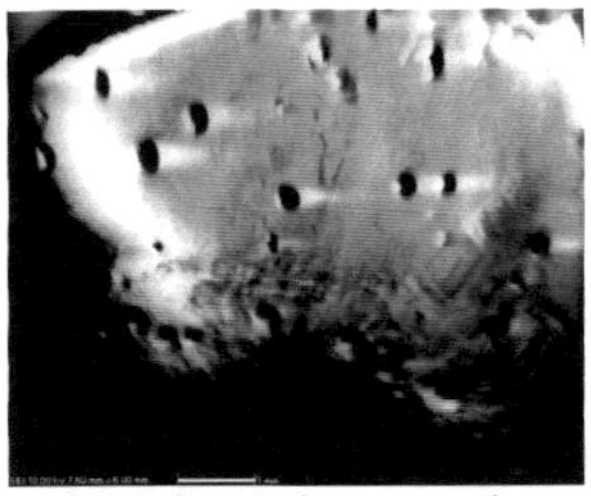

SEM for Ashory modern

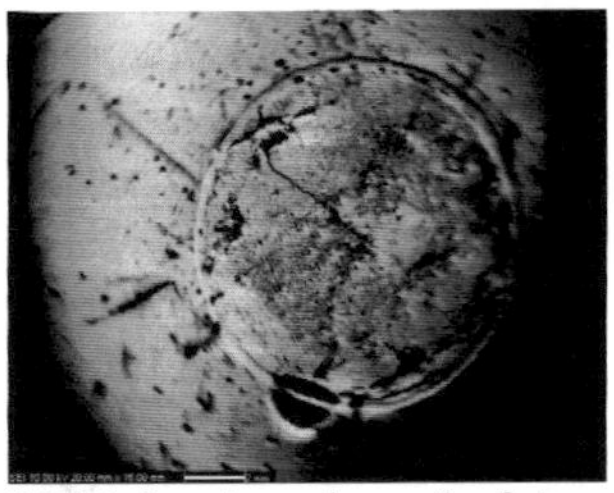

SEM for ring-shaped of Acady

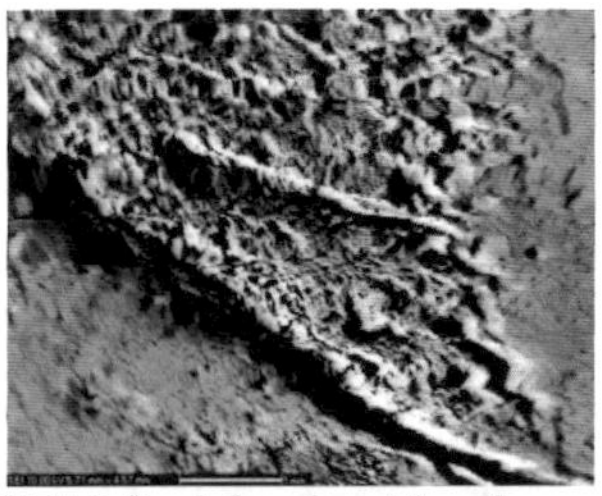

SEM for Islamic, Al-unbar

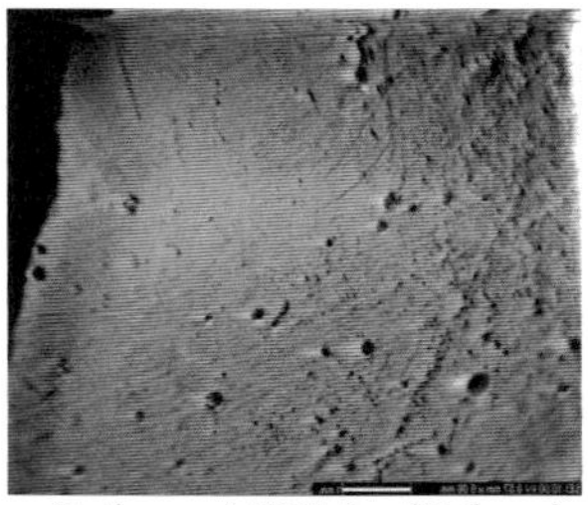
Enlarged SEM of Islamic

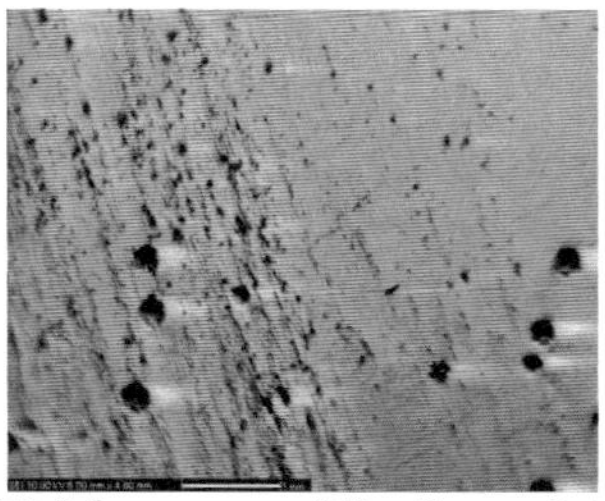
SEM for pores of Surface of Acady

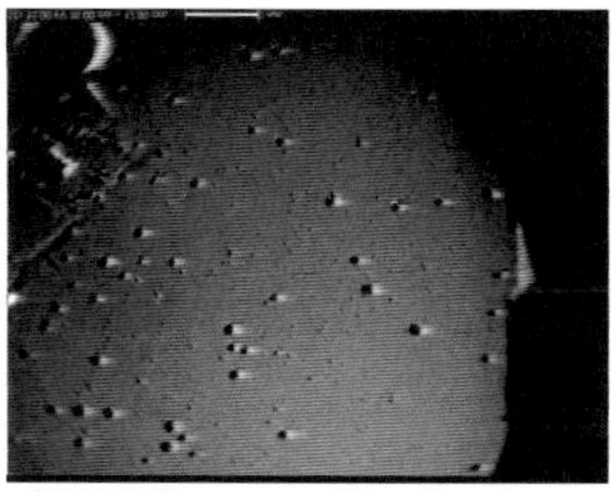
SEM for pore distribution

Figure (6): Scanning electron micrographs for gold-coated surfaces for some of the studied samples.

4. Conclusions

From the results obtained, the following conclusions could be mentioned:

4.1. The heat treatment temperature was sufficient and appropriate and may reached up to 1200 °C, and although these objects contained relatively high proportions of silica and alumina, however high densification occurred, as indicated by the rounded closed pores.

4.2. The fact that the values of pore size ranging from 250-300 microns lead to the conclusion that the particle size of mixtures used in the manufacture of these bodies were not very fine and the mixing procedures were good enough.

4.3. Sintering times used appeared to be enough and there were no signs that the length of the sintering periods was more than that required since their internal structures appeared to be homogenous and did not contain any large merged pores, but; instead pores with converging sizes.

4.4. The different civilizations of ancient Iraq mostly used similar methods of forming and sintering to certain extent, and the difference was in the different content of elements of the clays used and specifically the particular content of sodium, calcium, sulfur, iron, aluminum and magnesium.

4.5. PIXE & XRF are non destructive, rapid, sensitive and multi-elemental technique, and they can complement each other sometimes. Some elements could be seen better with XRF and vice versa.

4.6. Some potters in earlier civilizations may put brand or some signs making their product recognizable as it was noted by-chance under SEM- microscope on the surface of Acadian sample, and this still needs to be studied more deeply.

References

[1] Lipinski, E. (ed.), State and Temple Economy in the Ancient Near East, 1. Leuven: OLAS (1979): 257-409
[2] Jacobs, L. Causes for the pale colour of second millennium BC pottery from several sites in Mesopotamia. In Mery, S. (ed.), Science
[3] Jean-Claude Dran, Accelerators in Art and Archaeology, proceedings of EPAC 2002, Paris, France
[4] M. Roumié, B. Nsouli, K. Zahraman, A. Reslan, Nuclear Instruments and Methods in Physics Research, Section B: Volumes 219–220, June 2004, Pages389-393
[5] M. Roumié, M. Hageali, K. Zahraman, B. Nsouli, G. Younes, Nuclear Instruments and Methods in Physics Research Section B: Volumes 219–220, June 2004, Pages 871-874
[6] M. Roumié, S. Y. Waksman, B. Nsouli, P. Reynolds, S. Lemaître., Nucl. Instr. and Meth. B 215 (2004) 196

IOP Conf. Series: Materials Science and Engineering **37** (2012) 012018 doi:10.1088/1757-899X/37/1/012018

[7] H.S. Cheng, Z.Q. Zhang, J. Song, M.H. Gao, D. Zhu, J.W. Lin, S.L. Feng, Nucl. Inst. and Meth. B 249 (2006) 601
[8] M. Roumié, P. Reynolds, C. Atallah, E. Bakraji, K. Zahraman, B. Nsouli, Nucl. Inst. and Meth. B 249 (2006) 612
[9] M. Roumié, B. Nsouli, K. Zahraman, A. Reslan, Nucl. Instr. and Meth. B 219-220 (2004) 389
[10] Ebeling, E. and Calmeyer, P. Gefab. RlA 3 (1957-1971): 182-3
[11] D.T. Potts, Mesopotamian Civilzation, 1986

AUTHOR INDEX

Institute of Physics
Dirac House, Temple Back
Bristol BS1 6BE UK

CURRAN ASSOCIATES INC
proceedings.com